Worlds Without End

Exoplanets, Habitability, and the Future of Humanity

Chris Impey

The MIT Press

Cambridge, Massachusetts | London, England

The MIT Press would like to thank the anonymous peer reviewers who provided comments on drafts of this book. The generous work of academic experts is essential for establishing the authority and quality of our publications. We acknowledge with gratitude the contributions of these otherwise uncredited readers.

This book was set in Stone Serif and Stone Sans by Westchester Publishing Services. Printed and bound in the United States of America.

Library of Congress Cataloging-in-Publication Data

Names: Impey, Chris, author.
Title: Worlds without end : exoplanets, habitability, and the future of
    humanity / Chris Impey.
Description: Cambridge, Massachusetts : The MIT Press, [2023] | Includes
    bibliographical references and index.
Identifiers: LCCN 2022018595 (print) | LCCN 2022018596 (ebook) |
    ISBN 9780262047661 (hardcover) | ISBN 9780262373074 (pdf) |
    ISBN 9780262373081 (epub)
Subjects: LCSH: Extrasolar planets. | Habitable planets. | Outer
    space—Exploration.
Classification: LCC QB820 .I47 2023 (print) | LCC QB820 (ebook) |
    DDC 523.2/4—dc23/eng/20220713
LC record available at https://lccn.loc.gov/2022018595
LC ebook record available at https://lccn.loc.gov/2022018596

10 9 8 7 6 5 4 3 2 1

Contents

Prologue:
The Best of All Possible Worlds

We have only ever known one world. Yet many others are orbiting distant suns. To recognize the potential for biology on the myriad of planets scattered throughout the vastness of space, consider the following environments, which at first seem hostile to life.

The darkness is absolute, and the pressure suffocating. This far under the ocean, we must visualize the scene because no diving suit or bathysphere could protect us from being crushed to oblivion. Below us, a region of the seafloor is illuminated by a dim, red glow. The light is coming from a volcanic vent where superheated water is forced out from the mantle. Blobs of hot, molten rock are ejected along with the water and then drift down to the seafloor as lumps of black volcanic rock.

Scanning the horizon, we see a jagged panorama of volcanic mountains. The air is thin and cold, and a keen wind blows across the high plateau. There is no hint of moisture in the air, and the ground underfoot crunches like sugar. The plain is white with streaks of yellow, orange, and red, with a coarse texture composed of dried salts and crystals. The air has an acrid edge to it. It has not rained here for a hundred years.

A tomb of rock. Not a chamber or underground structure, but a solid and almost seamless mass of granite. It's laced with fissures and crevasses, the result of cycles of heating and cooling over the eons. Water oozes down the rock surfaces, sticky with dissolved chemicals. No light can penetrate this far underground, hundreds of meters below the surface. The only energy source is a feeble flux of radiation from radioactive decay within the rocks. It feels claustrophobic, desolate, and uninhabitable.

A hundred meters from the cave entrance, little light can get through. Boiling water bubbles up from an underground source and forms steaming puddles on the cave floor. The cave walls glisten in the gloom. They have a sheen from toxic metals dissolved in the water—cadmium, lead, mercury, and arsenic. Vapor is like a suffocating shroud in the cave. It is silent except for the bubbling of the chemical cauldron along with the slow and steady plink of drips from the vaulted ceiling overhead.

The sky is ocher red, and darkened by smoke and soot. There is no oxygen; its place has been taken by acrid methane and ammonia. Every few hours, a meteor hurtles into view and slams into the ground, throwing up a cloud of volcanic rock and ash. On this young planet, active volcanoes dot the horizon, and the ground is still soft underfoot. Magma has recently solidified, and oceans have not yet condensed from steam. There is a constant seismic trembling underfoot.

Are any of these forbidding, austere environments host to life? All of them. Where in the universe are they located? Here, on the home planet at different times in its history.

Three hundred years ago, German polymath Gottfried Wilhelm Leibniz argued that ours was the best of all possible worlds. He was trying to solve the conundrum of why an all-powerful God would allow evil to exist. Even if God had created many worlds, Leibniz contended, this perfection and actions based on reason mean we inhabit the best of all possible worlds.[1] Voltaire lampooned the idea in his novella *Candide*, where Dr. Pangloss, who is a parody of Leibnitz, clings to this conclusion even as catastrophes rain down on him.[2]

Now that we know there are at least as many exoplanets as stars, we can put our planet in a larger context. The conclusion: Earth is not the best of all possible worlds. Life did not start in naturalist Charles Darwin's "warm little pond." It probably began in the darkness and crushing pressure of a hydrothermal vent.[3] Early humans did not aggregate in the balmy tropics. They roamed the arid expanses of Africa's deserts, and hunted near the glaciers and on the tundra of the last Ice Age. Life is tenacious and has gripped this planet like a fever for four billion years. Life is adaptable and has radiated into almost every conceivable environmental niche.

The universe contains an immense number of exoplanets. These worlds without end are of interest to planetary scientists and geologists, but that is not our main concern. We are fixed on their potential to host biology. We are curious whether the experiment that began on Earth soon after its formation has been replicated anywhere else. For if we live in a biological universe, it's a universe much more fascinating than one completely described by the laws of physics. The universe becomes a place of almost infinite possibilities—a place where we might one day make another home or find companionship.

1
Searching for Distant Worlds

Thirty years ago, the Voyager space probe took a picture of our planet, looking back from six billion kilometers away. Earth was a tiny dot, less than 1 pixel of the 640,000 pixels in the image. Reflecting on the significance of the image, astronomer Carl Sagan called it the "pale blue dot." In 1994, he wrote, "That's home. That's us. On it everyone you love, everyone you know, everyone you ever heard of, every human being who ever was, lived out their lives. . . . The Earth is the only world known so far to harbor life. There is nowhere else, in the near future, to which our species could migrate. Visit, yes. Settle, not yet. Like it or not, for the moment the Earth is where we make our stand."[1]

Nearly thirty years ago, we started to discover planets beyond our solar system. First, it was a trickle and then a flood. The discovery of exoplanets—planets around other stars—has spawned an exciting new field of science in which we are encouraged to think creatively about habitability and consider Earth's place in the pantheon of planets. In these first few chapters, we learn about the many worlds that might harbor life. Who knows? One day in the far future, we might explore some of these planets, as we explore the planets of our solar system—and one of them could even provide a new home if and when Earth becomes uninhabitable.

1
The Visionaries

He was visualizing distant worlds and their possibilities long before astronomers were able to detect them. He wrote about travel to outer space and the settlement of remote planets. He described first encounters with alien life-forms. Some of the worlds had unusual atmospheres, including liquefied air, others had gigantic life-forms, and one operated under alternative physical laws. He imagined these worlds as populated by humans and even by their machines or robots.

The first-known musings about other planets do not come from the science fiction of the twentieth century. They are from a writer who lived during the height of the Roman Empire, nearly two thousand years ago.

Lucian of Samasota was born in 125 CE, near the Euphrates River at the far eastern edge of the Roman Empire. His family was lower middle class, and he was apprenticed to his uncle, who owned a statue-making shop. Lucian proved to have little talent as a sculptor, but was a gifted writer and orator, so he became a traveling lecturer, visiting universities throughout the Roman Empire. He lived in Athens for a decade, where he was a celebrity, and later in his life, gained wealth as a government official in Egypt. Lucian was one of the most important philosophers of his age. Due to his satirical and ironic style, and desire to be entertaining, it is difficult to know how seriously to take his statements. He frequently ridiculed religious practices, superstition, and belief in the paranormal. He championed society free from social, racial, and economic distinctions.

Lucian had an enormous impact on Western literature. His ideas influenced the work of William Shakespeare, Desiderius Erasmus, Jonathan Swift, Voltaire, Johann Wolfgang Goethe, Denis Diderot, Cyrano de Bergerac, and many others reaching into the modern era.[1]

Lucian's book *A True Story* is arguably the world's first novel, a precursor to the seminal science fiction of Jules Verne and H. G. Wells.[2] But it is more than that. It's a series of inventive allegories, satirical commentary on people who claim to be experts, and parody of the myths of his time. Writer and critic Kingsley Amis said, "The sprightliness and sophistication of *A True Story* make it read like a joke at the expense of all early-modern science fiction."[3]

The novel starts by saying that nothing in it is true, and that everything in it is a complete and utter lie. Lucian and his fellow travelers are carried to the Moon by a whirlwind, and become enmeshed in a war between the king of the Moon and king of the Sun. Both of their armies are populated by bizarre, hybrid life-forms. They then visit a diverse set of metaphoric worlds, populated by shape-shifting women, robots, and warring giants. One of these worlds is populated by a nonmaterial race functioning under a dream logic, which allows Lucian to explore a set of alternate physical laws. Even now, few people would speculate so boldly about life beyond Earth.

The tradition of thinking expansively about realms beyond Earth dates to the sixth century BCE. The word "world" comes from Saxon tribes in the early Middle Ages, with a rough meaning "age of man." It refers to this planet or the material universe. The corresponding word in Latin is borrowed from the Greek "cosmos." As first used by Pythagoras, cosmos conveys a view of the universe as a complex yet orderly system. The creation of the universe establishes order out of chaos.

In the beginning, the debate was purely philosophical. Anaximander took infinity as a foundational concept, and argued that innumerable worlds, or cosmic systems, are born and destroyed. Anaximenes extended this to the simultaneous multiplicity of worlds.

In the mind of Democritus, these innumerable worlds were random aggregations of simple atoms. Some of these worlds were like our own universe, while others were completely different. In the third century BCE, Epicurus wrote, "There is an infinite number of worlds, similar to ours, and an infinite number of different worlds. . . . [O]ne must agree that in all these worlds, without any exception, there are animals and plants and all of the living things we observe."[4]

Around this time, Greek philosophers asserted that Earth was round and had estimated its size—and Greek philosopher Aristarchus had even suggested that our planet orbits the Sun. But Aristarchus's Sun-centered model was squashed by the weight of arguments made by the most influential philosopher of all time: Aristotle.

To Aristotle, it was obvious that Earth is stationary since we don't feel any motion. The Sun, Moon, and planets all appear to move around us, so Earth must be at the center of the universe. And if there were more than one world or center, objects in space would have no natural way to move. He was emphatic that Earth stood alone: "The world must be unique. . . . [T]here cannot be several worlds."[5] The weight of Aristotle's intellect and reputation smothered the idea of many worlds, with the brief exception of Lucian's fantastic work of fiction, for nearly two millennia.[6]

My narrative takes place within the Western tradition, but Eastern systems of thought have long acknowledged the idea of many worlds. I've been traveling to India for over a decade to teach Buddhist monks and nuns cosmology. I teach the view of the universe that emerged with the Greek philosophers, and then took its modern scientific form through the theories of physicists Isaac Newton and Albert Einstein. But rich conversations with these monastics have enlightened me about the long Buddhist tradition of belief in a multitude of worlds, perhaps infinite in number, that occupy a vast and ancient universe. Humans are by no means at the pinnacle of the sentient creatures that inhabit these worlds. It's a strikingly modern vision.[7]

My account of exoplanet history is familiar to Western readers, but it's not the only story that could be told. There's a direct path from the development of the logical tools of the scientific method by ancient Greeks to the revolution in physics that culminates with Newton. From there, the Industrial Revolution leverages physics to build machines that do work and that in turn leads to Western dominance of the world economic order.[8] Newton codified gravity, optics, and mechanics, and his insights are still used by astronomers when they detect exoplanets and entrepreneurs when they launch rockets into Earth orbit.

The next visionary of worlds beyond Earth working in the Christian tradition was born to a family of modest means in a small town near Naples in 1548. Drawn to a life of learning, Giordano Bruno entered the Dominican order and became a priest when he was twenty-four. Within three years, he had scandalized his order with controversial views on the Trinity. Excommunicated, he was a fugitive for fifteen years, traveling from town to town across Europe.[9] Bruno accepted the then-heretical heliocentric (Sun-centered) model of the universe that Nicolaus Copernicus had put forward in the 1540s—and he recognized that it opened the possibility that the universe might be enormous or even infinite. He was the first person to suppose that stars are like the Sun: they create their own light and heat, and they're orbited by other Earths. He speculated boldly that these other worlds are homes to animals and other inhabitants. All of these ideas are strikingly modern for their time— and came before anyone had ever gazed at the night sky through a telescope.

Bruno had the unfortunate habit of not suffering fools gladly and being intemperate in his criticism of people who disagreed with his views. He returned to Italy in 1591 and within months had been denounced to the Inquisition. He was extradited from Venice to Rome and endured a six-year trial, at the end of which he was presented with a list of eight heretical propositions that he had

Figure 1.1

The trial of Giordano Bruno by the Roman Inquisition, in a bronze relief by Ettore Ferrari, Campo de' Fiori, Rome, Italy. Bruno was convicted of heresy and put to death in 1600. (Credit: Jastrow, 2006, https://commons.wikimedia.org/wiki/File:Relief_Bruno_Campo_dei_Fiori_n1.jpg.)

published. Bruno angrily rejected the charges and maintained his belief in the plurality of worlds. In January 1600, all of his works were put on the Index of Prohibited Books, and he was remanded to the Roman authorities as "an impenitent, pertinacious, and obstinate heretic."[10]

Three weeks later, he was led at dawn to the Campo di Fiori, stripped naked, a metal spike was driven through his tongue to silence him, and he was burned at the stake.

Bruno was a philosopher and mystic, so his speculations had no evidence behind them, but soon after his death, the scientific basis for planets was established. Johannes Kepler made the breakthrough that showed Earth is no different than other planets in their motions through space. The week Bruno was executed in Rome, Kepler met the meticulous astronomer Tycho Brahe in Prague and realized that Brahe had accumulated a vast set of data on the motion of Mars. Over the next nine years, while working as imperial mathematician to the Holy Roman Emperor, Kepler deduced the first two of his three laws of planetary motion. He published the work in 1609, noting in the introduction that "I prove philosophically not only that Earth is round, not only that it is contemptibly small, but also that it is carried along among the stars."[11]

The next year, Galileo Galilei electrified the scientific world with his observations using the newly invented telescope. *Starry Messenger* was the pamphlet he wrote about the discoveries.[12] His sketches of the Moon showed that it was a geological world like Earth, with mountains, plains, and valleys. By tracking four dots of light that appeared near Jupiter, he showed that they were moons too, orbiting the giant planet. These observations were ultimately a death knell for the idea that Earth was singular and the center of the universe. Galileo counted ten times more stars with his telescope than could be seen with the naked eye, and the gauzy haze of the Milky Way resolved into the light of a myriad of faint stars. Galileo's support of the Copernican model led to his trial with the Catholic Church, but unlike Bruno, he was lucky to pay with nine years of house arrest rather than his life.

I led an astronomical tour of Italy some years ago and got a sense of the high stakes associated with these new ideas in astronomy. A statue of Bruno looms over the market stalls in the Campo de' Fiori in Rome. I tried to visualize the scene when he was paraded around the square, naked in a donkey cart, and then burned at the stake. Farther north, in Venice, I climbed to the top of the campanile in St. Mark's Square where Galileo demonstrated his telescope to the doge of Venice in 1609. It's said that some people disbelieved the evidence of their eyes because it conflicted with the geocentric dogma of the time. Then I visited the villa just outside Florence where Galileo lived from 1631 until he died in 1642. This elegant house on a hill, nicknamed "the Jewel," looks out across vineyards and farm-land. Surrounded by poplars and the sounds of birdsong, with the summer heat cooled by a breeze, it occurred to me that there were worse places to be confined.

Kepler was quick to endorse Galileo's observations, but was dis-appointed that Galileo never sent him a version of his telescope or commented on Kepler's book. In 1611, he circulated the manuscript of a book that would eventually be published after his death, called *Somnium*. The title is the Latin word for dream. *Somnium* is an auda-cious piece of science as well as a seminal work of science fiction.[13] It began as a student thesis that aimed to describe how Earth looks as seen from the Moon. Kepler explored the change in perspective enabled by the Copernican Revolution, where Earth was not the cen-ter of the universe. He accurately described the motions of the plan-ets and stars as seen from the Moon, and speculated meaningfully about the difficulty of space travel along with aspects of lunar geog-raphy, geology, and even biology. He added the dream framework of a trip to the Moon and inserted autobiographical details. *Som-nium* carries an echo of Lucian's much earlier book.

During the long time when scientists could only speculate about planets orbiting other stars, science fiction offered the canvas on which to paint visions of travel to other worlds and the forms of life that might exist there.

The earliest writing about epic adventures and imaginary places is more appropriately called fantasy. In a fantasy, the setting of a fictional universe is inspired by real-world myths and folklore, and science themes are usually absent. There is an overlap, however, between the genres. The four-thousand-year-old *Epic of Gilgamesh* is the oldest-surviving work of literature, and science fiction writer Lester del Rey argued that "science fiction is precisely as old as the first recorded fiction. That is the Epic of Gilgamesh."[14] Other early texts that involved travel to other worlds include the *Maha-bharata* (ninth century BCE), *Ramayana* (fifth century BCE), plays of Aristophanes (fourth century BCE), and several stories from *One Thousand and One Nights* (eighth century CE).

Modern science fiction evolved during the seventeenth through the nineteenth centuries, with the Industrial Revolution and major discoveries in astronomy, physics, and mathematics. It emerged as a medium to explore the relationship between technology, society, and the individual. Science fiction records the history of our growing understanding of the universe and position of our species in it.[15] The first popular success in the science fiction genre was Mary Shelley's *Frankenstein* in 1818. Her short novel features the danger of a "mad scientist" experimenting with advanced technology.

At the end of the nineteenth century, Verne and Wells broadened the genre and encouraged many imitators. Verne mixed romantic adventures with state-of-the-art technology that he logically extrapolated into the future. His stories were optimistic and found a large audience, making him the second most translated author in the world, after Shakespeare.[16] One of his most successful books was *From the Earth to the Moon*, part of the long lineage of lunar exploration fantasies that starts with Lucian of Samasota. Verne uses plausible physics to fire a projectile containing three men to the Moon. The book was published almost exactly a century before Apollo 8 made the fantasy into fact. Wells was a prolific English writer, whose range spanned novels and short stories, social commentary, history, satire, and biography. He took science fiction from pure adventure to the realm of ideas. As a futurist, he anticipated aircraft, space travel, nuclear weapons, satellite television, and even the World Wide Web and Wikipedia. He wrote about a trip to the Moon in *The First Men in the Moon* and Martians in *The War of the Worlds*. The latter is the precursor to dozens of books and films about alien invasions. Wells had a pessimistic view of the future, but was unrivaled in the way he anticipated the power of science to change the world.

Another key figure in the development of modern science fiction was Hugo Gernsback, who started the magazine *Amazing Stories* in

1926. The "golden age" of science fiction lasted several decades.[17] Sci-fi has diversified into many subgenres, from cyberpunk to space opera, and has even achieved some of the literary respect that long eluded it.

Yet the diversification of genres obscures the fact that science fiction has been a Western monoculture. The protagonists are almost exclusively white and male. Time is linear, and there's generally an unbridled confidence that technological progress is a force for good.[18] Recently, these boundaries have been broken, perhaps most notably in science fiction inspired by African culture, and told from a female perspective, such as that by Nnedi Okorafor and N. K. Jemisin.[19]

In science fiction and fantasy from the last century, we see precursors of concern over our impact on Earth as well as speculation on what might await us on other worlds. The "hard" science fiction that most strongly anticipates exoplanets is the set of books that look at the future of humans off Earth as they explore the galaxy and encounter other civilizations. Early examples were *Triplanetary* from 1934, the first book in the Lensman series by E. E. "Doc" Smith, and *Foundation* from 1942, the first volume of a trilogy by Isaac Asimov. Hard science fiction bends, but never breaks, the known laws of physics. It uses remote planets as stages and backdrops for stories, living worlds, and places we might one day visit and call home. Starting in the 1950s, science fiction found a wider audience, thanks to blockbuster movies and popular television series.

Our increasingly sophisticated understanding of the cosmos helped fuel the rise of science fiction. By the late 1990s, we knew a lot about the universe. The big bang model was affirmed by detailed observations of the "cosmic microwave background"—radiation left over from the universe's early hot phase. Cosmologists had ascertained that the universe is 13.8 billion years old, and its two main ingredients are dark energy and dark matter.[20] The former causes the cosmic expansion to accelerate, and the latter holds galaxies together and permeates the space between them.[21] Images from the

Hubble Space Telescope were used to count galaxies through space and back to the dawn of time. The census found 100 million or so galaxies. Multiplying by the average number of stars per galaxies yields 10^{24} stars in the observable universe. That is a truly incredible million billion billion stars.[22] Astronomers supposed that exoplanets must exist, but they had not been able to detect planets around any of the vast numbers of stars. The only planets whose existence was known were the eight in our solar system.[23]

It was like we had counted all the adults in the world but didn't yet know if any of them had children. That was about to change.

2
Doppler Wobble

*The stillness of the night was broken by the sound of corks popping
and birdsong as the eastern sky began to pale. Early on July 5, 1995,
at the Haute Provence Observatory in southern France, Swiss astronomers Michel Mayor and Didier Queloz had woken their wives to
celebrate with champagne and raspberry tarts. They had found an
invisible object about half the mass of Jupiter in a rapid orbit around
a bright star in the Pegasus constellation. The duo had first observed
51 Pegasi in 1994, but they chose to wait until the following observing season to be sure of their result.[1] Staring at the data, there was
no doubt. It was the first planet ever discovered around another star
like the Sun.*

*The planet was fifty light-years away, pretty much in our back-
yard. Why did it take so long to find it, and how was it done using a
telescope not big enough to be in the top seventy worldwide?*

The answer to the question of why it took astronomers so long to
find the first exoplanet lies in the fact that planets are dim compared to the stars around which they orbit (their "host" stars).
Consider trying to observe a planet several light-years away using
a telescope. Planets don't emit light; they reflect light produced by
their host star. This is true of the planets in our solar system (and of
course, the Moon); we only see them because they reflect sunlight.

A planet intercepts a tiny fraction of the total light output of the
star. Jupiter, for example, is 140 thousand kilometers in diameter,
but it orbits 780 million kilometers from the Sun. As a result, Jupiter
catches less than one-millionth of 1 percent of the light the Sun
gives out. To make things worse, no planet reflects all the light that

falls on it. The proportion of reflected light is a planet's "albedo"; Jupiter's albedo is 0.52 (it reflects 52 percent of the light that falls on it), while Earth's is just 0.3. To have the best chance of seeing a dim planet next to its bright host star, we need to catch it when it appears as far from the star's bright disk as possible (at "elongation") rather than in front of or behind it, for example (at "conjunction"). If you do this, you will be looking side-on—but then only half the reflected light comes our way.

The upshot of all of this is that from a viewpoint far away in deep space, Jupiter would appear less than one-billionth as bright as the Sun. That's like trying to see the feeble glow of a firefly next to the bright glare of a lighthouse beacon.

Given these enormous challenges standing in the way of the direct observation of exoplanets, it's perhaps no surprise that the first successful detection of a planet around a star other than the Sun was made using an indirect method—a method that relies on the fact that a star "wobbles" as a planet orbits it.

As described by Newton's law of gravitation, any two objects in space are pulled toward each other with a force. The strength of the force depends on the masses of the two objects and distance between them. Effectively, any planet tugs on a star with the same force as the star tugs on the planet. As a result, the star does not remain stationary; it wobbles to and fro as the planet revolves around it. The star and planet orbit a common center of gravity. The magnitude of the wobble depends on the mass of the star and planet along with the orbital distance. The speed of the Sun's wobble caused by the orbit of Jupiter is 12 meters per second or 27 miles per hour (mph)—sprinting speed.

Astronomers determine a star's wobble by observing the "Doppler shift" it imprints on the star's light. We're familiar with the Doppler effect with sound, in which a siren's pitch is higher as it approaches and lower as it recedes. If any source of waves is moving toward you, the waves are squashed or "blueshifted"; if the source is moving away from you, they're stretched or "redshifted." The size of this effect is the ratio of the speed of the motion to the speed of the waves. Light speed is 300,000 kilometers per second, so the Doppler shift in the light of a star like the Sun caused by the presence of a planet like Jupiter is only 0.000003%. That sounds desperate, but it's thirty times better than detecting the one part in a billion reflected light.

The small wobble a planet causes in the star it orbits, called a "reflex motion," was the key to the discovery of exoplanets. Spectroscopy was the tool.[2] Astronomers often spread starlight into a spectrum; a smooth light distribution is overlaid by narrow dark features imprinted in cooler gas at the edge of the star. In the 1930s, these spectral features were used to learn what stars are made of, and how they create energy by nuclear fusion. For exoplanet detection, the role of the spectral lines is to act as a set of wavelength reference markers. The wavelengths of the lines in the star's spectrum are compared to lines from the same elements in the lab, and the Doppler shift gives the velocity of the star. If the star is being tugged backward and forward by an orbiting planet, the lines shift redward (the wavelengths are lengthened) when the star is moving away from us and blueward (the wavelengths are shortened) when it is moving toward us. The shifts repeat in a cycle each time the planet completes an orbit.

Imagine a nearby sunlike star orbited by a Jupiter-like planet. Over the orbit, the Doppler shift appears as a periodic variation (the redshifts and blueshifts vary smoothly like a sine wave). The amplitude of the sine wave (the size of the wavelength shift) allows astronomers to work out the velocity shift, which in turn gives the mass of the planet.

What is required is patience—and fortitude because exoplanet hunting was a field of broken dreams for much of the twentieth century. Reputations were tarnished, if not destroyed. Claims of a dark body orbiting within the 70 Ophiuchi system were made three times in the first half of the century, but were all later shown to

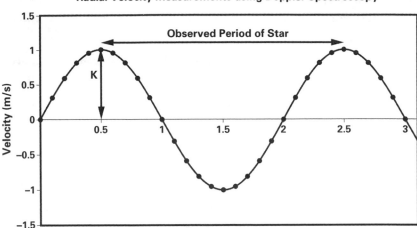

Figure 2.1

The radial velocity of a star measured with spectroscopy, showing a periodic Doppler variation caused by a planet. In this case, the period is two years, and the amplitude, giving the mass of the planet, is one meter per second. (Credit: Cosmogoblin, 2020, https://commons.wikimedia.org/wiki/File:Doppler_Shift_vs_Time.svg.)

be in error. In the 1960s, two planets were thought to be orbiting Barnard's star. That also didn't pan out. In 1991, astronomers at the University of Manchester claimed they had detected a planet in orbit around a pulsar (a spinning neutron star)—but then retracted the claim six months later. The titles of the two papers tell the story: "A Planet Orbiting the Neutron Star PSR 1829–10" was published in *Nature* in July and then "No Planet Orbiting PSR 1829–10" appeared in *Nature* six months later.[3]

Advances in experimental techniques opened the door to success. Sensitive electronic detectors replaced photographic plates, and Roger Griffin, working at the University of Cambridge, showed how to use all the lines in a stellar spectrum rather than just one for radial velocity measurement. This improved the accuracy by combining information from hundreds of spectral lines. At the back end of the telescope, optical fibers were used to take the light to a spectrograph on the floor of the dome, improving measurement stability and accuracy.

Another key innovation was putting a gas absorption cell in the beam of the telescope.[4] The spectrum of the gas vapor in the absorption cell is imprinted on the starlight, and acts like a reference spectrum or measuring stick for the Doppler shift. All of these improvements in technology reduced velocity errors by a factor of a hundred in just a decade—from a thousand kilometers per second to ten kilometers per second. Because the Doppler signal repeats with the period of the planet's orbit, data covering many orbits could be combined to further beat down the errors.

Yet exoplanets remained elusive. In the 1980s, Gordon Walker and Bruce Campbell at the University of British Columbia did an extensive search for substellar companions orbiting sunlike stars. They saw evidence for a Jupiter-mass planet orbiting the bright star Gamma Cephei and published the result in 1988. But they decided the planet signal might have been coming from the star, so they retracted the claim in 1992. They even got poisoned by their instrument when the corrosive and dangerous hydrogen fluoride they used in their absorption cell leaked out.

Meanwhile, Geoff Marcy and Paul Butler at the University of California also worked for years and came up empty-handed. With the experience of the Canadians as a cautionary tale, Butler built an absorption cell using the innocuous gas iodine. These groups competed on experimental techniques, but were patient because Jupiter takes twelve years to orbit the Sun, so finding Jupiter-like planets around other stars would take years or decades.

By 1994, Mayor had built a new precision spectrograph in Geneva and was ready for a challenge: "When we have a new instrument like this, we say, okay, what can we do best? It was just the level to do a planet search! We were driven by technology."[5]

Finding exoplanets requires patience, fortitude, and luck. When Mayor and Queloz, his graduate student, started their search, they came from a background of studying binary stars. These stars had orbital periods of days or weeks, so they were looking for binaries where one of the objects was dim or had low mass, and thus was potentially a planet. With observations over a week in 1994, Queloz saw an intriguing signal in the unassuming star fifty light-years away called 51 Pegasi (in the constellation Pegasus), often shortened to "51 Peg." He sent a telex to his adviser: "Michel, I think I have found a planet." But Mayor was extremely cautious. Partly, this was because of the history of failed searches and false detections. It was also because the planet orbiting 51 Peg was no ordinary one. It was half the mass of Jupiter and moving on a blistering four-day orbit. It revolved around its host star twenty times faster than Mercury orbits the Sun (thousands of times faster than Jupiter) and at a distance one-eighth as far as Mercury is from the Sun (less than one-hundredth the distance Jupiter is from the Sun).

Nervous about rushing into print, Mayor and Queloz waited almost a year until 51 Peg was visible in the next observing season. With the new data, there was no doubt—and that was when they popped the champagne cork. Their discovery paper was still being reviewed when they presented the data at a conference in Florence a few months later. News spread like wildfire. Astronomers reacted with a mixture of excitement and skepticism.[6] Four days later and five thousand miles away, Marcy and Butler had time already scheduled at Lick Observatory in California, and they quickly confirmed the Swiss team's discovery. Finding the first exoplanet didn't need a large telescope because it was revealed by precise data on its parent star; 51 Peg is bright enough to be seen with the naked eye.

A bitter irony for the California team is that 51 Peg was on its observing list, but the team had not reduced the data, in part because computers were painfully slow back then, yet mostly because everyone anticipated that giant planets would have multiyear orbits. The news must also have been frustrating for Walker and Campbell. They had published the first exoplanet discovery in 1988, but withdrew the claim in 1992. Better data in 2003, however, showed that there *was* an exoplanet orbiting Gamma Cephei. Walker and Campbell were wrong to retract; they had made the first discovery! This episode demonstrates the messiness of scientific research, where detections are made at the limits of the data and certainty is elusive.

The first exoplanet discovery was front-page news worldwide. With the realization that giant planets could have rapid orbits, Marcy and Butler started grinding through their backlog of data. It took two computers all day to reduce a single night of 51 Peg data. Digging through eight years of data took twenty computers nine months. Butler's eye was caught by the Doppler curve for 70 Virginis. It indicated a planet seven times the mass of Jupiter orbiting its bright star every four months.[7] Here is Butler describing the moment he realized he was looking at the second exoplanet to ever be discovered: "After nine years of working toward this moment, I was stunned, silent. . . . In the absolute quiet of a New Year's Eve Sunday morning, I sat for the next hour looking at the signal. I had the sense that Johannes Kepler was standing over my shoulder, looking at the same signal."[8]

The floodgates opened. Astronomers mined the data they had already gathered and soon found more exoplanets. By 1997, the total was ten. By the end of 2002, it had risen to a hundred.

Nearly all the exoplanets in this first tranche of discoveries were alike—ideally suited to detection by analyzing the Doppler curve: hot Jupiter-like planets on tight orbits around sunlike stars. These planets were puzzling because the standard theory of planet formation holds that there's too little material close to a star to form a gas giant planet. Theorists had anticipated that planets might migrate inward after formation, but the prediction was overlooked. They had concluded, for example, that Jupiter was not born where it is today.[9] It is now generally accepted that hot Jupiters formed at large distances and migrated inward.[10] Another concern was that our solar system architecture might not be typical. Nevertheless, astronomers were cautious about drawing that conclusion since the Doppler method will most easily find planets that are massive, giving a bigger signal, and on shorter orbits, where less data is needed. They knew they would have to take data on many stars for long time spans to see the true population of exoplanets.

In 1999, the first multiple-planet system was found: three massive planets orbiting the star Upsilon Andromedae, in the constellation of Andromeda.[11] The radial velocity method (the Doppler wobble approach, described above) can detect more than one planet using a single Doppler curve because each planet contributes a unique velocity signature. The contributions of planets with different masses and orbits add linearly in the data, just as the various modes of vibration (harmonics) of a guitar string contribute to the guitar's overall sound. This may sound easy in principle, but in practice it's tricky because the data is often "noisy," which can suggest several different multiplanet solutions.[12] To overcome this, Doppler planet hunters gather data for longer, over more complete orbits, allowing them to "clean" the data of noise.

Astronomers soon chafed at pedestrian exoplanet names. The planet orbiting 51 Peg was dubbed Bellerophon by Marcy, after the ancient Greek hero who rode the winged horse Pegasus. In 2014, the International Astronomical Union announced a system for naming prominent exoplanets based on nominations and votes from the public.[13] The winning name was Dimidium, Latin for half, referring to the planet's mass of half that of Jupiter. Upsilon Andromedae's three planets were named Saffar, Sahm, and Majriti, after noted Arab astronomers from the tenth century. But unofficial names were more fun. Debra Fischer, one of the few women working in a male-dominated field and codiscoverer of the Upsilon Andromedae planets, preferred the names proposed by a fourth-grade class: Ginky for the smallest,

Twopiter for the one that was twice Jupiter's mass, and Fourpiter for the one that was four times the mass of Jupiter. It even caught on at a conference where she gave a talk. A prominent theorist was trying to describe the instability of the system. "Well, yes, if B—I mean, C," he said, struggling for a naming convention. Then he threw up his hands and continued: "Look, if Twopiter gets too close to Fourpiter, it's all over and Ginky will be ejected from the system."[14]

The culmination of the story of the discovery of the first exoplanet came twenty-four years after Mayor and Queloz sipped champagne in the south of France. In December 2019, they traveled to Stockholm to receive the Nobel Prize in Physics.[15] Science is not about prizes, and scientists are not supposed to be driven by that kind of recognition, but most researchers are competitive, and a Nobel Prize is the pinnacle of scientific achievement. It's even more unusual for astronomers to win in a category that's devoted to physics.[16]

Two researchers who won the race but not the Nobel Prize, in addition to Walker and Campbell, are Aleksander Wolszczan and Dale Frail. Early in 1992—three years before the discovery of 51 Peg was confirmed—they announced the discovery of two planets orbiting the pulsar PSR 1257+12.[17] This claim was confirmed later with better data—and a third planet was detected. Two of these exoplanets are a few times Earth's mass, and one is a fiftieth of Earth's mass and only twice the mass of the Moon, making it by far the lowest-mass exoplanet ever found. Pulsars are spinning neutron stars left behind after a supernova explosion. Because these planets formed unusually and were not orbiting a normal star like the Sun, this discovery was unjustly neglected.

After decades of failed searches and retracted claims by other astronomers, success was sweet. Mayor had not spent his career looking for exoplanets but rather was in the right place with the right instrument when the opportunity arose. In another twist of fate, Queloz was only at the telescope for the discovery because

Mayor's most senior graduate student had died in a car accident a few months earlier. The California team was in the hunt yet just missed being first.[18] Mayor is refreshingly levelheaded about the accolades: "If you consider colleagues doing science in many different fields, they are much better physicists. It's not about the quality of the people. We have to keep our feet on the ground."[19]

With these discoveries, centuries of speculation ended. Our solar system was not unique, and an exciting new field of science was born.

3
Chasing Shadows

"Look, Bill, everyone agrees that what you're doing is not going to work." This verdict was delivered by the new chief of the Space Science Division at NASA Ames.[1] Bill Borucki was used to adversity. He had pitched the mission in 1992, 1994, 1996, and 1998, and each time it had been rejected. Ten years and many technical reviews and anxious moments later, NASA launched the Kepler space telescope on a Delta II rocket from Cape Canaveral in Florida.[2] As Kepler was launched, 365 exoplanets were known. By the time its work was wrapped up a decade later, it had found 2,600 new exoplanets and 3,600 exoplanet candidates.

The road to Kepler's discoveries began in a parking lot at the High-Altitude Observatory in Boulder, Colorado. A young grad student from Canada, David Charbonneau, had cobbled together a four-inch telescope from cheap parts. He was chasing shadows.

At the time of the first exoplanet discoveries, astronomers already knew there was another way they might be detected. It involves measuring the variation in brightness of the light a star emits. If a planetary system is oriented just so, the planet will cross the face of the star briefly. This is called a transit or eclipse. If we were observing our solar system from afar and Jupiter passed in front of the Sun, it would block some of the light and dim the Sun briefly. The dimming would be slight—only 1 percent—and only last for just over a day during each twelve-year orbit.[3] Measuring the brightness of light is called "photometry," and a photometric chart representing the brightness over time of the star is called the "light curve." Photometry is the key to the success of the transit method.

To observe a transit, the relative orientation of the planet's orbit is crucial. If you observe from a point "above" the orbital plane, so you can see the whole orbit, the planet will never pass between you and the star. If you view it from side-on, however, you will observe the full transit. In between these two extremes are other orientations in which the planet either does or does not pass in front of the star from your observation point. The odds are considerably better in systems in which a planet orbits close to the star—and as explored in the previous chapter, many of the first exoplanets were hot Jupiters orbiting close to their stars. Proximity raises the odds considerably.[4] That's the promise that drew David Charbonneau, a young Harvard graduate student, away from his initial project in theoretical cosmology toward the hot new field of exoplanets. To test the idea and technology, Charbonneau used data about an exoplanet that two teams had already detected orbiting a sunlike star called HD 209458.

Exoplanet research was marked by intense competition—and occasional collaboration. The Swiss team members (Michel Mayor and Didier Queloz) bagged the first exoplanet, and then the California group (Geoff Marcy and Paul Butler) mined their data and found the next two, as we saw in the last chapter. Both groups worked flat out to improve their instruments. When they each got tantalizing data on HD 209458, 160 light-years away, they pooled their efforts and were rewarded by the detection of a planet two-thirds the mass of Jupiter on a dizzying 3.5-day orbit. The combined data allowed a prediction of the timing of transits in the system. This was the data Charbonneau used to become the first to detect the transit of an exoplanet.

At the end of August 1999, Charbonneau turned his tiny home-built telescope toward HD 209458. A light-sensitive charge-coupled device (CCD) detector attached to it measured the amount of the star's light reaching the telescope. Charbonneau watched intently as the light from the star dipped by 1.5 percent for three hours at the predicted time and then rose to its original level. When he saw the same dip a week (two orbits) later, he knew he had nailed the first exoplanet transit. Charbonneau only made it by a whisker; a competing group made the same observation two months later.[5] He was hooked. Charbonneau is now a Harvard professor and exoplanet pundit with many career awards. Otherwise, he is a typical Canadian who brews beer and cider, and coaches his four daughters on their respective hockey teams.[6]

That first photometric transit detection was pivotal. It confirmed a discovery that had been made using the radial velocity (Doppler wobble) method, proving a second, independent way to discover exoplanets. What's more, the transit added crucial information. Doppler detection yielded the planet mass and orbital period. Transit detection yielded the planet size and orbital period. Combining the mass and size gave a new piece of information: the average density. This planet is 30 percent larger than Jupiter, but two-thirds of its mass. It is an unusually "puffy" gas giant; astronomers still debate why it has such low density. Proximity to its host star means its temperature is a blistering 1400 K (the Kelvin temperature scale is used in this book, and for high temperatures, it's close to degrees in Celsius).

The next landmark in chasing shadows came in 2002. Transits last a small fraction of the orbital time and only occur with a favorable orientation of the observer to the orbit, so discovering exoplanets through transits is like looking for a needle in a haystack. The best way to catch them is to carry out a "survey"—that is, look at lots of stars simultaneously for a long time. A group at Harvard found the first planet to be discovered by its transits by sifting through candidates from a microlensing survey. Microlensing (see chapter 5) is the temporary brightening, rather than dimming, of a star due to the passage of another star in front of it.[7] After sifting through dozens of candidates and tossing out binary star "imposters," only one had the right transit signature. It was a faint star five thousand light-years away, twenty times more distant than any other star previously found to host an exoplanet. The exoplanet was extreme: a Jupiter-size world whipping around its star every twenty-nine hours and baked to a temperature of 2000 K—hot enough to vaporize iron.[8]

To see how the field exploded in the past decades, I return to Borucki and the story of the Kepler space telescope, which took the search for exoplanets by the transit method to dizzying heights. Borucki spent his first ten years at NASA designing heat shields for the Apollo program. He was intrigued by a paper proposing that exoplanets could be detected by transits of their parent stars. In the mid-1980s, he organized two workshops to examine the technology needed to detect transits.[9] He deduced that simultaneous measurements of at least ten thousand stars would allow the detection of Jupiter-size transits. But Bill knew that the big hook for exoplanets was detecting habitable planets like Earth, not gas giants like Jupiter.[10] The detection of Earth-size transits needed observations from space because the precision required to detect that more subtle transit can only be done in the stable space environment. Let's see why.

Observing Jupiter from afar, from a position that allows you to see its orbit perfectly side-on, you would see each transit dim the Sun's light by 1 percent for twenty-nine hours, as noted above—or 1/100,000 of the twelve-year orbital period. Only one in a thousand randomly oriented planetary systems will be aligned well enough to make the transit visible. So even if every solar system has a Jupiter-like planet, the Kepler experiment is like staring at ten thousand 100-watt light bulbs for twelve years and hoping to see ten of them dim to 99 watts for a little more than one day sometime during the entire survey. The situation is different for earthlike planets. When Earth crosses the Sun as seen from afar, it dims the Sun by just 0.01 percent for thirteen hours. Although this is shorter than twenty-nine hours, Earth's orbital period is one year rather than twelve, so thirteen hours represents 1/700 of the orbital time—a much larger fraction. And because Earth is so much closer to the Sun than Jupiter, one in two hundred randomly oriented solar systems will be aligned to make the transit visible, compared with one in a thousand. Let's say every planetary system out there has an earthlike planet. If we stare at those same ten thousand light bulbs for one year, fifty of them will dim for half a day sometime during the year—much better odds. But the dimming will only be from 100 watts to 99.99 watts—an extraordinary challenge to detect. That kind of precision can only be achieved by observing from space.

A survey for transiting exoplanets overcomes the low probability of detection by looking at a huge number of stars at the same time, but there's another problem: false detections. Stars have intrinsic variations that can masquerade as exoplanets. Binary stars eclipse each other and that can also mimic an exoplanet. And there are two types of stars that are approximately the same size as gas giant planets: white dwarfs and brown dwarfs. Light curves can't tell these apart because they only indicate the size of the transiting object. Where possible, Doppler data is needed to show that the object is in the planet mass range, below thirteen times the mass of Jupiter. This is why the Kepler results have so many "exoplanet candidates" as well as a large number of confirmed exoplanets: the candidates have yet to be confirmed by other methods.

Borucki beat down the technical challenges one by one and assembled a team. When NASA announced a new set of missions that would answer questions about our solar system and might address exoplanets, Borucki saw his opportunity. So began a series of four proposals and rejections through the 1990s. While Borucki was trying to get the idea approved, he repeatedly asked his team members if they still believed in the project or if they wanted out. Everyone stayed in—and Carl Sagan's opinion was particularly important. Sagan, who had a towering scientific reputation, was the most famous astronomer in the world at that time. He was unwavering in his support. Sagan said, "I know you failed—keep trying."[11]

The new chief of NASA's Space Science Division—the one who told Borucki that no one thought the idea would work—was Dave

Morrison. Luckily, Morrison didn't close the door entirely. He convened an external committee to take a fresh look at the project, led by Jill Tarter from the SETI Institute (see chapter 17). If the committee was not convinced, Borucki would terminate the project. After a daylong presentation, the committee agreed the project might be successful. Several members of the committee were so enthusiastic they asked to join the team. Funds were released, and the project was given the green light.

In December 2001, Kepler was selected as a NASA Discovery-class space mission. Much work lay ahead before the launch, but during the intervening time the first exoplanets had been discovered, transits had been detected from the ground, and the motivation for the mission was stronger than ever. Still, the challenges that faced Borucki and his team were illustrated by the smaller CoRoT satellite, the first space mission designed to detect transiting exoplanets. Launched in 2006 by the European Space Agency (ESA), the satellite only found thirty-four exoplanets during six years of operation.[12] This modest yield was due to the failure of one of the satellite's two detectors and the fact that its telescope mirror was only eleven inches in diameter, smaller than the telescopes used by many amateur astronomers.

The mirror of Kepler's telescope is only one meter in diameter—so small it would not make the top two hundred largest telescopes.[13] Borucki called it the most boring mission ever, built to take a picture of a single patch of sky every six seconds for years. What did the mission accomplish?[14] The first several hundred exoplanets were found with the Doppler wobble method, but Kepler became the workhorse for exoplanet detection. While the Hubble Space Telescope is like a Swiss Army Knife, packed with many different tools for observing the sky, Kepler was built to do just one thing, but do it exceedingly well. It stared at one 115-square-degree patch of the sky next to the Cygnus constellation, taking a CCD image every six seconds, gathering light curves. There were a half-million stars in its field of view, and 150,000 were sunlike stars and so were selected for observation. In such a small patch of sky, to get enough stars meant going faint, so most of them were hundreds of light-years away. That's still in our galactic "backyard," yet it doesn't include the stars closest to the Sun. A transit had to be seen consecutive times to be confirmed. During the nearly ten years of its time in orbit, Kepler found over twenty-six hundred exoplanets, including several hundred that

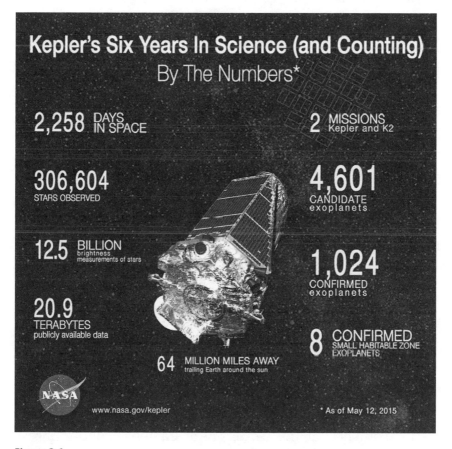

Figure 3.1

Statistics from the six core years of the Kepler space telescope, showing the amount of data taken, number of individual measurements, and yield of confirmed and candidate exoplanets. (Credit: NASA Ames/W. Stenzel, "Kepler's Six Years in Science (and Counting)," last updated 2017, https://www.nasa.gov/ames/kepler/six-years-in-science/.)

are close to the size of Earth.[15] Fifty are in the habitable zones of their stars, at a distance where liquid water can exist on the surface of the planet.

For the first time in history, we learned that there are other worlds with the potential to host life.

Kepler revealed to us that small worlds are common. The galaxy is teeming with terrestrial-size planets. As many as half of all stars have rocky planets like Earth, with the possibility of life on any of them. It turns out that planets are diverse. The most common planet size detected—between the size of Earth and Neptune—doesn't even exist in our solar system. Planetary systems are diverse too. Kepler discovered many multiplanet systems—one with eight planets, as many as in our solar system. A lot of the systems are compact, with planets orbiting much closer to their stars than Mercury does to the Sun. There's much to learn about how such compact systems form and evolve. Most profoundly, the Kepler data showed that there are more planets than stars in our galaxy. That projects to several hundred billion throughout the Milky Way, and a similar number in each of the hundreds of billions of other galaxies in the universe.[16] The total is a staggering number.

Resourceful NASA scientists were able to squeeze even more exoplanets out of Kepler after the end of the primary mission. For three and a half years after its launch, the spacecraft stared at the same 150,000 stars. But in 2013, the second of its four stabilizing reaction wheels failed. Over the next four years, in what was called the K2 mission, the spacecraft surveyed multiple target regions, repositioning itself several times a day to counteract the pressure from solar radiation. This noisy data was challenging to analyze until researchers at the University of California at Los Angeles developed algorithms to pluck light curves from the noise. These researchers cleverly estimated the "false positive" rate by injecting inverted light curves into the data, where the transit caused the star to brighten instead of dim. The result was nearly 750 new exoplanets, with 90 percent reliability.[17]

The Kepler spacecraft was deactivated with a "good night" command on November 15, 2018, the 388th anniversary of Johannes Kepler's death. Borucki was forty-four when he started thinking about chasing shadows, sixty-two when the Kepler mission was finally selected, and seventy-nine when the telescope was retired. He recalled his excitement before the launch: "When I was a kid, I built rockets. And I had the privilege of going up to the big rocket booster, which was fifteen stories high, and I thought, this is going to carry my experiment into space. It was a really special experience."[18]

When Borucki retired in 2015, he handed over the reins as mission scientist to Natalie Batalha. She joined the Kepler team when she was just thirty and then became leader of a half-billion-dollar project. Managing a team of hundreds of scientists is a challenge; big egos and sharp elbows are not in short supply. As Batalha frames it, "Science is hard, people are harder." Astronomy is transitioning to a new generation, where women are better represented and more prominent. Batalha started as a business major at Berkeley, but switched to physics when she encountered the elegance of applying mathematics to nature. She says, "I was moved that the universe could be described in numbers." Now when she runs in the foothills near her home in northern California, she thinks of how Kepler has brought the stars to life. Her subject resonates: "No longer mere observers of the void, we become pilgrims of the Milky Way."[19]

Batalha takes pleasure in summing up the exotic worlds that Kepler revealed: "We have found lava worlds with one hemisphere that's an ocean—an ocean not of water but of molten rock. We have found disintegrating planets literally breaking up before our eyes, because of orbits close to the parent star. We have found planets that are orbiting not one, but two stars. We find planets as old as the galaxy itself; I think that was a huge surprise. It means that the raw materials for planet formation are available in the earliest stages of a galaxy's life. We find planets associated with dead stars, orbiting white dwarfs. We also find interesting architectures: compact systems, packed so tightly that planets feel gravitational interactions from one another."[20]

Kepler was phenomenally successful, but to find so many exoplanets it had to cast a net over many stars in a small patch of sky. That means the stars are relatively faint and distant. Its closest discovery is the unusual Kepler-444 system, 119 light-years away and 11 billion years old. Five rocky planets between the masses of Mercury and Venus are on tight orbits of the system's brightest star, which has two red dwarf companions.[21] Over 90 percent of Kepler exoplanets are more than 1,000 light-years away, thereby making follow-up observations challenging.[22]

As part of that follow-up to Kepler, NASA launched the Transiting Exoplanet Survey Satellite (TESS) in 2018 (for more details, see chapter 6). TESS is surveying two hundred thousand bright stars across the entire sky, anticipating the discovery of three hundred Earths and super-Earths.[23] Its planets will be around one-tenth as far and a hundred times brighter than Kepler's planets, making detailed observations from ground-based telescopes much easier. The results from TESS are still being analyzed, but it is already delivering the goods; a batch of twenty-two hundred exoplanet candidates was announced in early 2021, a number that had risen to over four thousand by mid-2022.[24]

People say science is expensive. The movie *Avatar* cost $250 million. For a quarter of a billion dollars, you can get a movie by a famous director about life on a distant world. But for twice that, you can get a mission that will actually discover hundreds of habitable earthlike planets. Science is a pretty good deal.

4

Seeing Is Believing

*The indirect methods of exoplanet detection I've explored so far—
Doppler spectroscopy and transit photometry—are fiendishly clever,
and have given birth to a new field of astronomy.*

*A tug here, a shadow there, and eventually you have a catalog of
thousands of exoplanets. But there is something unsatisfying about
this type of evidence. Why can't astronomers just take a picture?
After all, seeing is believing.*

There are two main obstacles to capturing images of exoplanets. First,
there is an enormous difference between the amount of light we
receive from a planet and the amount of light we receive from a star.
A planet is a rock surrounded by gas (in our solar system, Mercury
is an exception). As noted in chapter 2, a planet doesn't emit any
light, so we only see it by the light it reflects from its host star—and
planets are small and far from their stars, so they only intercept and
reflect a tiny fraction of the starlight. In our solar system, that fraction is about one-billionth for Jupiter and one-tenth-billionth for
Earth. Fireflies use about 100 microwatts (10^{-4} W) for flashing, and a
large football or baseball stadium uses about a megawatt (10^6 W) of
power.[1] If the Sun is the sum of all the stadium floodlights, Earth is
one firefly and Jupiter is a huddle of ten fireflies.

The other challenge is caused by the fact that the average distance between stars is millions of times larger than the typical distance
between a star and its planets. When we look at a nearby planetary
system, all the planets appear close to the star. For our solar system

observed from a distance of 33 light-years (10 parsecs), the angular separation between Jupiter and the Sun is 0.5 arc seconds—one arc second is 1/3600 of a degree—and between Earth and the Sun, it's 0.1 arc seconds.

These are minuscule angles. If you looked at a ruler two hundred meters away, the Jupiter-Sun separation would be the distance between millimeter tick marks on the ruler, and the Earth-Sun separation would be one-fifth as wide. That's not much more than the width of a human hair at that distance—an almost imperceptible angle.[2]

From the above, the odds of success are best when the exoplanet is large and far from the glare of the star. So imaging was targeted initially to gas giant planets with large orbits and long orbital periods. It also helps to observe in the infrared region of the electromagnetic spectrum. Planets are warm and so they emit infrared radiation—and while stars are even warmer, the intensity of infrared relative to the visible and ultraviolet radiation they emit is smaller. In other words, a curve showing intensity versus wavelength falls away in the infrared region of the spectrum for stars, but the opposite is true for planets. As a result, observing a planetary system in longer wavelengths (infrared) provides a far greater contrast between exoplanet and star than observing at shorter (visible, ultraviolet) wavelengths. The factor of a billion to one for a Jupiter-like planet relative to a sunlike star in visible light improves tenfold, to a hundred million to one at 5 microns (5×10^{-6} m), a wavelength ten times longer than visible light. The factor of ten billion to one for an earthlike planet relative to a sunlike star improves a hundredfold to a hundred million to one at 5 microns—an even greater gain.[3] Young planets are hotter than older ones, so astronomers try to target systems where a sunlike star is less than a billion years old.

The first exoplanet to be discovered by direct imaging was seen in 2004.[4] As a sign of the difficulty of this method, over 150 exoplanets had already been found by the radial velocity or transit methods. The contrast was favorable because the star was a brown dwarf—a dim and low-mass type of star not hot enough to fuse hydrogen into helium. Brown dwarfs are often and more appropriately referred to as failed stars. The companion is estimated to be three to ten times the mass of Jupiter, and it's young and still glowing red-hot. Over the next few billion years, it will cool and shrink, ending up slightly smaller than Jupiter.

Despite these challenges, fifty exoplanets have been imaged. That's a tiny fraction of the over five thousand that have been discovered, but it's still impressive. Unsurprisingly, most are large—five to ten times the mass of Jupiter—and much farther from their stars than Jupiter is from the Sun.[5] Hot Jupiters are not present in our solar system, and giant cold Jupiters, as some of these are, are unfamiliar too. The most impressive direct image is of a system with four massive planets, between six and nine times the mass of Jupiter, orbiting HR 8799—a star in the constellation Pegasus that's bright enough to see with the naked eye.[6] The star is thirty million years old, so these planets formed relatively recently. They're bracketed inside and outside by a disk of cold gas and dust left over from their formation. The lightest planet ever imaged is a super-Earth that orbits Proxima Centauri, a red dwarf star only 4.2 light-years away. Proxima Centauri, in the constellation Centaurus, is the closest star apart from our own Sun. This planet is tantalizingly close—but unfortunately, so frigid that it's unlikely to be habitable.[7]

For the reasons set out above, imaging has only been successful for extreme exoplanets: those that are nearby, massive, and on large orbits. It's also effective when the planet is young and hot from its recent formation. You might think that more powerful telescopes, with high magnifications, would be able to separate the two images. After all, a large and powerful telescope can produce far more detailed images than a cheap pair of binoculars with low magnification. Unfortunately, it's not that simple. The sharpness of an image made with any optical system—such as a telescope, microscope, or

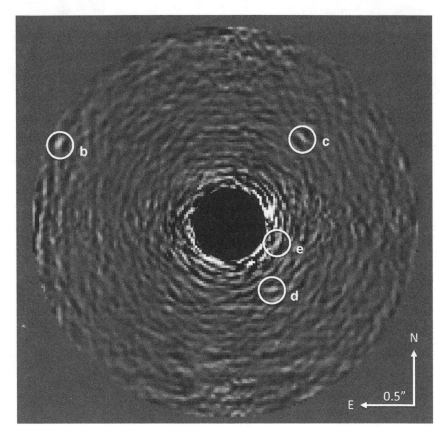

Figure 4.1
Image of the four exoplanets orbiting the star HR8799, taken with the five-meter Hale Telescope at Palomar Observatory. The light from the star has been suppressed. These gas giant planets are ten to seventy times the Earth-Sun distance from their star. (Credit: Project 1640, in "Project 1640 Reveals Precise Composition Information about Four HR 8799 Planets" by W. Clavin, 2013, *SciTechDaily*, https://scitechdaily.com /project-1640-reveals-precise-composition-information-about-four-hr-8799-planets/.)

camera—is subject to a fundamental limit. This limit has nothing to do with the quality of the mirror or lens; it is due to the wave nature of light. As light (or any other wave) passes an edge, it spreads out—a phenomenon called diffraction. The light from distant stars and planets spreads over a significant angle; most of it ends up in a central disk, and the rest in a series of concentric rings. This characteristic of light was first described fully by English astronomer

George Airy.[8] The diffraction limit defines the minimum angle for separating two objects in the sky.[9] It depends on the wavelength of the light and size of the telescope mirror (or lens aperture). The bigger the telescope and shorter the light's wavelength, the sharper the images. But even a perfect optical system can't turn a point source of light into an image that's a point. In other words, it doesn't matter what the contrast is or how high the magnification of the telescope is: planets can't be seen if the starlight and reflected light from the planet are blended.

What does this mean in practice? For visible light, the diffraction limit of the human eye, a small optical device, is 60 arc seconds or 1 arc minute (1/60 of a degree).[10] The Moon is 30 arc minutes across, so you can see or resolve features about 1/30 of the Moon's diameter, allowing you to make out large craters and mountain ranges. The larger the mirror or aperture of the optical system, the sharper the images. The diffraction limit of a small 10-centimeter (4-inch) telescope is 1 arc second, for a 1-meter telescope it's 0.1 arc seconds, and for NASA's 2.4-meter Hubble Space Telescope it's 0.05 arc seconds. Using the examples given earlier, when looking at a planetary system 33 light-years away, we should be able to separate a Jupiter-like planet from its sunlike star using a 20-centimeter telescope (diffraction limit 0.5 arc second), and an earthlike planet with a 1-meter telescope (diffraction limit 0.1 arc second). But there's a catch!

Earth's atmosphere messes up the calculation completely. Looking into space through the atmosphere is like looking through a fogged and uneven pane of glass—worse still, one that is constantly moving. The churning motions of moist air and dust particles distort and obscure starlight that passes through the atmosphere. What would otherwise be a sharp image is instead jumbled up and smeared out by turbulence, which astronomers call the "seeing."[11] Seeing at most sea level locations is 1 to 1.5 arc seconds (this is the amount of the blurring caused by the turbulence), while the best mountaintop observatories have seeing of 0.7 to 0.8 arc seconds.

When the atmospheric conditions are stable, the seeing at altitude can be as low as 0.4 arc seconds.[12] Since 0.4 arc seconds is the diffraction limit of a 25-centimeter telescope, the potential of larger-aperture telescopes to make sharper images is completely wasted. All of their images are blurred by Earth's atmosphere. This blurring is fatal to most attempts to make images of exoplanets.

An easy yet expensive solution to the problem is to put the telescope in orbit, above Earth's turbulent atmosphere. The Hubble Space Telescope has been astronomy's premier facility in part because of its location (that accolade has now passed to the James Webb Space Telescope [JWST], which was launched in 2021). Hubble's mirror is not large by modern standards, but it makes ten times sharper images in space than it would on the ground.

Thanks to the ingenuity of telescope engineers, all is not lost for ground-based telescopes. Astronomers use clever methods to "cheat" the atmosphere and make sharper images. "Adaptive optics" is the term for actively altering the shape of a mirror to compensate for the distortions introduced by the atmosphere.[13] US astronomer Horace Babcock proposed adaptive optics in 1953; it took half a century to develop the technology and make it work.[14] (Babcock was prescient in another area too. His 1938 PhD dissertation contained evidence for dark matter in the Andromeda galaxy. But he didn't pursue it, so dark matter in galaxies didn't receive serious attention until decades later.)

You have a natural adaptive optics system in your head. The eye is a flexible device. It sends signals to your brain, your brain interprets the image, and then it applies corrections either voluntarily or involuntarily.[15] Muscles compress the soft lens to adjust the focus distance. There is a tilt capability to track moving objects. The iris opens and closes to control the light falling onto the retina. When you squint, you are applying an aperture stop, spatial filter, and phase control mechanism. You probably had no idea your eyes were so high tech!

Adaptive optics systems on large telescopes are sophisticated and cost millions of dollars. A small fraction of the incoming light is sent to a gizmo called a wave front sensor.[16] This sensor measures the amount of distortion across the image caused by the atmosphere. A fast computer calculates the mirror shape required to correct the distortions, and the surface of a deformable mirror is altered accordingly. The primary mirrors of most telescopes are too big and rigid to deform, so the correction is usually applied to the smaller secondary mirror. The atmosphere churns quickly, and all of these calculations and corrections must be done a hundred times a second. Since the target of the observation is often faint, the reference for the incoming wave front is taken from any nearby bright star. A more flexible solution, since there may not be a suitably bright star nearby, is to bounce a powerful laser off the atmosphere at altitudes of fifty to eighty thousand feet. You've probably seen pictures, evoking *Star Wars*, of telescopes shooting beams of laser light into the night.[17]

It took twenty years for these technologies to be perfected. Now every large telescope uses adaptive optics. I became familiar with adaptive optics when it was being rolled out on the telescopes I use in Arizona and Chile. Many nights were given over to teams of engineers to work the bugs out of the system, leading to tension with astronomers, who want every clear night to be used for astronomy. We don't begrudge engineering time since we all want telescopes to make the sharpest images possible. But I harbored a suspicion that the engineers were having so much fun, they didn't want to give the telescope up to the astronomers.

Adaptive optics sharpens images, so it helps with the separation issue, but it doesn't address the contrast challenge. Stars are billions of times brighter than their attendant planets. Can anything be done to reduce how much of the star's light a telescope receives? Yes. A coronagraph is a solid disk that blocks a star's light—although as you will see, it is not quite that simple.

Bernard Lyot was a coronagraph pioneer. Born in 1879 in Paris, his father died when he was seven, and he attended the evocatively named Graduate School of Electricity. He was lured into astronomy by the books of astronomer Camille Flammarion and was observing with his own telescope as a teenager. At twenty-seven, he got a job at the Meudon Observatory, where he rose to the rank of chief astronomer and worked the rest of his life.[18] Lyot became an expert on light and optical phenomena. He was obsessed with finding a new way to observe the Sun's corona, its extremely hot outer atmosphere. The corona is normally only visible during a solar eclipse when the Moon perfectly blocks the Sun's light. Lyot labored for a decade to perfect a coronagraph, an instrument to use with a telescope so that the corona could be seen without waiting for an eclipse. His coronagraph suppressed sunlight by a factor of a million. In 1939, he showed a spectacular movie of the Sun's shimmering corona at a meeting of the International Astronomical Union. He was honored with awards for that achievement. Lyot died in 1952 doing what he loved. He suffered a heart attack while returning from an eclipse expedition in Sudan.

Coronagraphs are used in the quest to capture images of exoplanets, and block out, or occult, the starlight and allow the feeble light reflected by a planet to pass unimpeded. It sounds like all you need to do is put a small opaque disk in the telescope path, but in practice, it's a lot more complicated. The occulting disk creates its own diffraction pattern, which must be suppressed—and light bouncing off the telescope structure adds scattered light, which must be modeled and subtracted. Coronagraphs are designed for a specific telescope and its optical characteristics.[19] A coronagraph in space is more stable and so performs better than a coronagraph on a ground-based telescope. The coronagraph on the Hubble Space Telescope, however, can only suppress starlight by a factor of a million at a separation of 0.5 arc seconds, which is far short of being able to image a Jupiter-like planet around a nearby star.[20] The technology development continues.

Another way to block starlight is to prevent it from ever reaching your telescope.[21] A "starshade" is a screen placed far in front of a telescope in space to block the star's light and leave the planet's reflected light to be caught by the telescope. Poetically, but for reasons that have to do with optics, it is shaped like a sunflower.

Astronomer Sara Seager is a starshade evangelist. She has been pushing starshades for a decade. Seager has a model of one petal hanging on the wall of her office at MIT. She takes another model on the road to classrooms and conferences—and the halls of Congress, trying to raise the larger sums of money needed to make it a reality.[22]

Starshades can achieve high contrasts with conventional telescope designs and without any active control of the optics. Contrast sufficient to detect analogs of Jupiter and even Earth may be possible. Seager and others are working on a NASA concept that is incredibly ambitious. A space telescope would have a starshade placed 40,000 kilometers (25,000 miles) in front of it. The two spacecraft would have to be aligned with an accuracy of 1 meter (3 feet). Imagine the starshade scaled down to the size of a drink coaster and the telescope to the size of a pencil eraser. They'd be separated by 100 kilometers (60 miles), and these two free-floating objects would have to be aligned within 2 millimeters (about one-twelfth of an inch).[23] Formation flying with this precision has never been achieved—and it's impossible to test starshades fully before the mission, on the ground, so the engineering must be masterful and get it right the first time.

Seager knows her quest can seem quixotic: "The search for life on other planets, it's so—I don't like to say it out loud—but it's kind of far-fetched, you know. We have to get really lucky for there even to be life on a planet that's orbiting a very nearby star that's just bright enough so that we can see it." But she's resolute and determined. "If I want the starshade to succeed, I have to mastermind it. The world sees me as the one who will find another Earth."[24]

That may seem like confidence bordering on arrogance, but Seager has earned it. She was a quiet, isolated child growing up in Toronto who blossomed when she discovered astronomy. As a young graduate student, she leaped into the dark by working on exoplanets when the early detections were still the subject of skepticism and scorn. She experienced pervasive and dispiriting sexism as a young woman in a male-dominated field.[25] Seager faced adversity. Friendless for most of her childhood, she met her perfect match and married. Then her husband got a rare form of cancer and died at age forty-seven, leaving her with two small children to care for while she struggled to build her career. Doggedness and talent saw her through. She was promoted to full professor at MIT, won a raft of national awards, and in 2013, was named a MacArthur "Genius Grant" Fellow. Thanks in significant part to her tenacity, we're well on the way to being able to take pictures of distant Earths.

Seeing is believing. It's proof that exoplanets exist, and the first step on the road to studying them in detail. Exoplanets in existing images are tiny—little more than a few bright pixels. Just as Earth was when observed from four billion miles away by the Voyager spacecraft, Sagan's pale blue dot.[26] The nearest stars are tens of trillions of miles away, so their pale blue dots will be faint indeed.

5

Calling Pandora

The landscape is otherworldly yet strangely familiar. The dense vegetation suggests a rain forest on Earth, but it clings to cliffs that rise vertiginously, and above them are rock masses that seem to float in the air. A gas giant planet looms in the sky, along with four other moons. The star is close enough to create eternal dusk. In the dim light on the surface, the greenery is streaked with electric blue, the trademark of bioluminescence. Viewed up close, the forest brims with exotic flora and fauna. Blue-skinned humanoids ride horselike hexapods in the air, and small animals try to evade viperwolves and an apex predator called the thanator. In the soil of this lush paradise, root systems form electrochemical networks that have become sentient, creating a primitive, moon-wide "brain." This is a mysterious world, primal and terrifying.[1]

James Cameron created an entire fictional universe for his 2009 science fiction movie Avatar. *Pandora is in the Alpha Centauri system. At a distance of 4.4 light-years, it's the closest stellar system to Earth.[2] But it's not a planet; it's a moon—an exomoon. Now that we've discovered exoplanets, can we detect the moons that might orbit them? And might life exist on exomoons?*

In Cameron's movie, Pandora is one of many moons of the gas giant Polyphemus, named after the man-eating cyclops described in Homer's *Odyssey*. Pandora is named after the first human woman in Greek mythology, who opened a jar and released the evils of humanity.[3]

There's a real moon called Pandora; it's the thirteenth-largest moon of Saturn, small enough to fit inside London or Los Angeles. The real Pandora is airless, pockmarked by craters, and not nearly as interesting as Cameron's fictional world. *Avatar*'s Pandora is slightly smaller and has weaker gravity than Earth. Mountains hang in midair buoyed on strong magnetic fields, created by an abundant superconducting mineral dubbed unobtanium. Its biosphere brims with bioluminescent diversity. The atmosphere has enough oxygen but too much carbon dioxide for humans to breathe, so the blue-skinned Na'vi have special organs to metabolize carbon dioxide. The science in *Avatar* is speculative and often fanciful, but the world Cameron conjures up is visually audacious. Many people were captivated, including the science writer for the *New York Times*: "It has recreated the heart of biology: the naked-heart-stopping wonder of really seeing the living world."[4]

Fanciful as Pandora may be, there's good reason to believe that many, if not most, exoplanets may have moons in orbit around them. In the search for habitable worlds far out in space, it might therefore be worth looking at not only planets but also moons. Earth is the only planet in our solar system within a traditionally defined habitable zone: the range of distances from the Sun where liquid water can exist on the surface of a planet. But if we assume that the only essential ingredients for biology are water, a local source of energy, and carbon-rich material, then life might exist in other locations in the outer solar system—and not only planets. Mars is too cold for surface water, but there's convincing radar evidence for subsurface aquifers.[5] The oceans of Jupiter's moon Europa are also a good prospect, along with Saturn's small moon Enceladus. Altogether there are a dozen moons in the cold depths of space where biology is conceivable, and the list even includes the dwarf planet Pluto, which is on average forty times as far from the Sun as Earth is (forty astronomical units, or AU).[6] Even this far from the Sun, water under a moon's surface can be kept from freezing by a combination of pressure from overlying rock, radioactive heating within the moon's rock, and tidal squeezing by the planet it orbits. Moons boost our solar system's potentially habitable real estate substantially, and there is no reason to expect that it would be any different in other planetary systems.

Let's look at the challenge in detecting exomoons. The largest moons in our solar system—Jupiter's Ganymede and Saturn's Titan—are less than half the size of Earth and slightly smaller than Mars. Both are larger than Mercury. The next largest are Jupiter's Callisto, Io, and Europa, and then Neptune's Triton, all of which are larger than Pluto.[7] There's little overlap between the sizes of the biggest moons and the smallest terrestrial planets in our solar system.

The Doppler method has mostly discovered planets that range from Neptune up to Jupiter's mass. The lowest mass so far, found by Michel Mayor and his group, is a hot, rocky planet 1.7 times the mass of Earth.[8] Detecting exomoons is unlikely using this technique because the Doppler shift of a stellar spectrum resulting from a planet would be the same whether or not the planet had orbiting moons. The transit method has greater sensitivity to small objects. The Kepler mission found over two hundred Earth-size planets along with thirty smaller than Earth. Two systems are particularly striking. Kepler-42 is a red dwarf 131 light-years from the Sun. It has three exoplanets, all of which are smaller than Earth in size and probably in mass. The smallest of them is the size of Mars.[9] Kepler-37 is a sunlike star 209 light-years away hosting three hot, rocky planets. One is smaller than Earth, and another is smaller than Mercury and slightly larger than Earth's Moon, making it the smallest exoplanet known.[10] We have seen that pulsars can have small planets, but they are extremely rare and experienced a supernova explosion, so you wouldn't wish their habitats on your worst enemy.[11]

Despite the challenges, astronomers have been hunting for exomoons for a decade. Early exoplanet researchers could write a paper and get a newspaper headline with a single exotic discovery. Now that exoplanets are released in batches of hundreds, though, getting noticed is harder. But the first confirmed exomoon would be a big prize.

So far, the story sounds familiar from the early years of exoplanets. Claims retracted. Discoveries not confirmed by other groups. Detections on the hairy edge of the noise. Most searches have delved deep into data from the Kepler satellite. Exomoons are so small that only an enormous one could make a detectable blip in the light trace of a star, but the way they tug on the exoplanet they orbit affects the timing of eclipses.[12] Astronomers at the University of Western Ontario used this "transit timing variation" to identify eight possible exomoons, but skeptical reanalysis left only two as viable candidates.[13] Pulling such tiny signals from noisy Kepler data is extremely difficult. To be sure they're not being fooled, astronomers ran simulations of billions of possible star-planet-moon combinations and compared them to actual Kepler data to look for a good match. This approach consumed five million hours of processing time on a powerful NASA supercomputer.[14]

It turns out there's another tool for finding exoplanets, and it's also relevant in the search for exomoons. This cunning method relies on a central prediction of Einstein's general theory of relativity: mass bends light. Einstein replaced Newton's concept of space and time as linear, independent entities with a theory that merged them into one entity, called space-time. His theory of gravity was geometric, and it emerged from the mathematics that mass could bend space-time.[15] In an analogy with optics, mass can bend, focus, and magnify light just like a lens can. The effect is called gravitational lensing. It's most familiar from beautiful Hubble Space Telescope images of distorted and magnified galaxies.[16] A foreground galaxy or cluster acts as the lens, and it magnifies and brightens the light of the background galaxies. If the alignment is right, a massive foreground galaxy bends light from the background galaxies to such an extent that we see multiple, distorted images of the background galaxies. A far less massive object like a star doesn't bend light much—the angle is a tiny fraction of an arc second. But the light is still magnified and brightened. That's called microlensing.[17]

Our story of how it was used to detect exoplanets—and is the sharpest tool in the search for exomoons—starts with a fourteen-year-old Polish boy, brimming with intelligence. He grew up in grim, postwar Warsaw, a city obliterated by the Nazis and subjected to Soviet leader Joseph Stalin's deadly purges. His name is Bodhan Paczynski. Young Paczynski was up way past his bedtime, observing an eclipsing binary star with a small telescope belonging to the Warsaw Observatory. The observatory, situated on the outskirts of town, was an abandoned property with half-destroyed buildings. Paczynski got there on the tram and then by walking four kilometers through a pine forest. There was no electricity, so he worked by candlelight. The data he took would become part of his first research paper, which was published when he was only eighteen. So began a lifelong infatuation with stars that whirl around each other in a dance where one crosses the face of the other.[18]

Paczynski's brilliance was recognized early on. His teachers helped him cross the Iron Curtain to do research. At age twenty-two, he spent a year at Lick Observatory in California—the first of many visits to the West. He was part of a succession of young astronomers from Poland who worked at Lick. These immigrants were popular because they were not only good but also cheap since Poland was in dire economic straits. Each fresh recruit was trained by their predecessor in the tricks of the trade: how to smash dry ice and pack a cold box, how to get data off punch tape, and how to survive weeks alone on a mountaintop observatory. Locals were perplexed by Paczynski's rapid, almost unintelligible English; he had learned the language by reading the highly technical *Astrophysical Journal*. Returning to Poland, he endured food shortages, married his childhood sweetheart, and got his PhD, finishing the calculations on a primitive vacuum tube computer.

Paczynski was insightful and voracious in his astronomical interests. In the 1970s, he solved five fundamental problems in stellar evolution. He wrote a prescient paper that predicted how binary star systems could be used to detect gravitational waves.[19] At thirty-six, he became the youngest-ever member of the Polish Academy of Sciences. He traveled abroad frequently since it was the only way he could get access to computers fast enough to do the calculations he needed for his work. Often, his family was essentially held hostage to ensure his return to Poland. In 1981, he was giving lectures at Caltech when martial law was declared in Poland. The regime started arresting and detaining intellectuals. Since his wife was with him on that trip, Paczynski decided to stay in the United States. He spent the rest of his career as a professor at Princeton University.

In the early 1980s, Paczynski recognized that the increasing size of CCD cameras, along with rapid advances in the computer processing of images, would allow searches for the rare situation where a star is lensed by an intervening star or planet. He coined the term "microlensing." Most astronomers doubted the effect would ever be seen. Undaunted, he set up a collaboration and network of small telescopes to monitor the brightness of millions of stars night after night. In 2003, two decades after Paczynski first conceived of the project, microlensing delivered its first exoplanet.[20]

Here's how it works. If one star is precisely in front of another star, the foreground star will magnify the light of the background star. It is a substantial effect; the brightening can be a factor of two or three. But stars are moving through space, so the perfect alignment is fortuitous and temporary. The brightening lasts a few weeks and then it fades over a few weeks, back to the normal level. Smaller and less massive objects magnify light by less and for a shorter time. For example, a planet five times Earth's mass would brighten a background star by 10 percent for about eight hours. The effects of the planet (and any moon, if present) and star combine. This is another needle-in-a-haystack experiment. The odds of a perfect alignment are tiny, so hundreds of thousands of stars must be watched to catch one rare event. When the slow, monthlong brightening caused by the foreground star begins, astronomers observe the star frequently

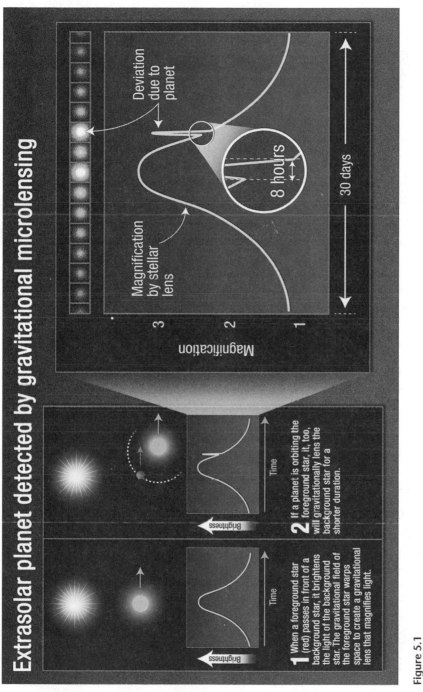

Figure 5.1

Microlensing is the brightening of a background star caused by the passage of a foreground star. If the star has a planet, it causes additional brightening, and if the planet has a moon (not shown) there will be even briefer extra brightening. (Credit: NASA, ESA, and K. Sahu/ STScI, "Extrasolar Planet Detected by Gravitational Microlensing," last updated 2021, https://exoplanets.nasa.gov/resources/53/extrasolar -planet-detected-by-gravitational-microlensing/.)

to try to catch the much-briefer brightening due to a planet. An orbiting exomoon would be an even smaller light blip on top of the slower magnifications, lasting just minutes. The moon is like a twig on a branch on a log in a vast forest.

As a graduate student, I had my own experience with looking for a needle in a haystack. My thesis work involved trying to find a rare type of active galaxy whose light output changed as the supermassive black hole that fueled the activity consumed nearby gas and stars. The active region is so small that the object looks like a faint star in the sky. In Australia, at the UK Schmidt Telescope, I took a series of photographic plates over weeks and months. There were hundreds of thousands of star images on each plate. Stars don't vary, so their brightness on each of the plates didn't change. But scattered across this region of the sky were a handful of active galaxies that were slightly brighter on some plates and slightly fainter on others. Hunting down supermassive black holes is exciting, but comparing millions of images by staring at them with a simple eyepiece felt like a Sisyphean task.

Microlensing has some great advantages for exoplanet hunting and a few downsides. It's sensitive to low mass planets and moons, and planets and moons orbiting at moderate to large distances from their stars. That makes it complementary to Doppler and transit searches, which work best at finding planets orbiting close to their stars. Microlensing can detect exoplanets at large distances, far across the Milky Way galaxy, and in principle even in other galaxies. If many thousands of stars are targeted and a microlensing event occurs in any of them, it will be detected. The big downside is that the event never repeats. With Doppler and transit methods, the orbit repeats, data can be rechecked, and the results from the two methods can be combined. With microlensing, the two stars are widely separated in space and on different trajectories. Like ships passing in the night, they'll never meet again.[21]

The method that Paczynski inspired has been a great success. Over a hundred exoplanets have been discovered in this way. Microlensing has found one of the lowest-mass exoplanets, just bigger than Earth's mass, and one of the most massive, thirteen times Jupiter's mass. It's the most sensitive method that astronomers possess for detecting dark objects. This latter object is so large that it straddles the boundary between planets and stars. It has even found one of the most robust candidates for an exomoon, making the fantasy of Pandora just a little more tangible.[22]

Paczynski's genius was realizing that small telescope networks could fill in a missing piece of astronomy: studying the behavior of variable objects, things that go bump in the night. "Most modern telescopes peer in great detail at tiny specks in the heavens," he said. "They can't keep track of what varies, pulsates or flares up and disappears forever."[23] He solved another variability puzzle: unexplained bursts of gamma rays discovered by satellites monitoring compliance with the Nuclear Test Ban Treaty. Almost alone, with most of his colleagues deeply skeptical, he argued that they originated far beyond the galaxy and were extremely luminous events. Paczynski was proved right in the 1990s by new satellite observations.

The father of microlensing was brilliant, but also plainspoken and modest. He worked tirelessly to support Polish astronomy after he emigrated, and almost single-handedly kept it going during the dark days of the Soviet regime.[24] In 2003, he was diagnosed with brain cancer. Confined to a wheelchair, he continued to work and mentor young scientists. He faced his disease with courage and humor, but the cancer would not relent. His was a great mind passing in the night, shining brightly and then fading.

6
The Next Wave

"What for me is absolutely magic, what is fascinating, is that from the first instrument implemented in 1977 we had 300 meters per second, and today, something like forty years later, we arrive at 0.1 meters per second. It's a factor of three thousand improvement, it's technological progress!"[1] Michel Mayor, discoverer of the first exoplanet, talks about the technology that led to his breakthrough. It's a reassuring cultural stereotype that a Swiss man studying the clockwork of planet orbits is a purveyor of precision.

The planet hunters didn't get smarter, they just developed better tools. They built spectrographs thousands of times more sensitive to detect tiny exoplanet motions. They developed CCDs to throw their net around thousands of times more stars simultaneously. They launched telescopes into space to detect photometric tremors far smaller than could be sensed by ground-based telescopes. In fields from genetics to physics to astronomy, science is driven by advances in technology.[2]

The pace of discovery has been frenetic. Exoplanet researchers wonder if they can keep up. "Is there a Moore's Law equivalent for exoplanets? Are we about to hit the exoplanet singularity?" So wondered Jessie Christiansen in 2016, when thirty-four hundred was the confirmed exoplanet count.[3] She grew up in rural Queensland, Australia, and went to an all-girls Catholic high school where she had no role models on how to be a scientist. "My favorite thing about space is all of the things we don't know yet," she says, even though she has been at the epicenter of exoplanet discovery as a research scientist at the NASA Exoplanet Science Institute.[4] Christiansen thinks the plethora of exoplanets is creating a confirmation bottleneck. For every confirmed planet, there are a half-dozen others that are just whispers in the data. Astronomers don't have the time or resources to investigate each one. "You have to prioritize," Christiansen admits, "You have to look at this list of planets and say, OK, which one do we really think we're going to learn the most about?" This means shifting to studying populations rather than individuals. "If it's the 80th hot Jupiter and we don't have any reason to believe it's going to be different from the 79 that came before it, are we going to scrutinize it in the same way we scrutinized the first 79?"[5]

The first quarter century of exoplanet discovery saw the body count double every two years. That's analogous to Moore's law, a foundational trend of our modern electronic life, where the former CEO of Intel described a doubling every two years of the number of transistors in an integrated circuit.[6] The search for exoplanets saw its first discovery in 1995; the count hit ten in 1998, a hundred in 2003, and a thousand in 2014. The trend is like Moore's law with a doubling in just over two years since 1995. In May 2022, the census stood at over five thousand, rising to over fourteen thousand if candidates are included.[7] After fifty years, the end of Moore's law for electronics is in sight, but exoplanets might well sustain their trend.[8]

Extrapolation is of course risky in science, but the capabilities that are arriving soon suggest a total of a hundred thousand by the end of this decade. Throwing all caution aside, 2050 could see a census of a hundred million exoplanets![9]

For the past decade, the heavy lifting of discovery was done by the Kepler spacecraft. What puts a gleam in the eyes of exoplanet hunters now—and as they contemplate the future?

In 2018, the baton passed from Kepler to TESS: NASA's Transiting Exoplanet Survey Satellite, a mission led by scientists at MIT. TESS

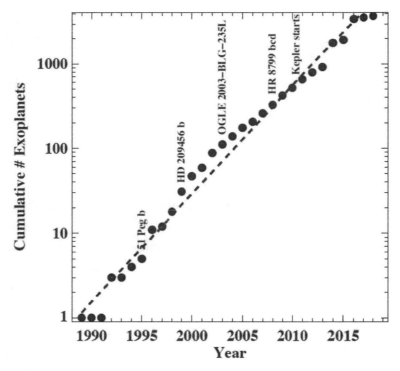

Figure 6.1
The cumulative number of exoplanets since the first discovery in 1995. The progression shows a doubling of the number in just over two years. (Credit: Data from the NASA Exoplanet Archive, image created by E. Mamajek, "Cumulative Number of Exoplanets Discoveries Versus Time," 2016, https://figshare.com/articles/figure/Cumulative_Number_of_Exoplanets_Discoveries_Versus_Time/4057704.)

got a funding assist from Google and launch assist from SpaceX, whose Falcon 9 rocket put it into a highly elliptical orbit.[10] It is the first mission ever to use this orbit, which mission lead scientist George Ricker called a "Goldilocks orbit," not too close to Earth and the Moon, yet not too far. The orbit is designed to balance out the gravitational tugs of Earth and the Moon, and so be stable for many years. At its high point, it's as far away as the Moon and so well above the hazards of the Van Allen radiation belts—a zone of energetic charged particles originating from the solar wind and captured by Earth's magnetic field. At its low point, it gets close enough to

Earth for data to be easily beamed down to the ground. This swoop-ing orbit lets TESS cover four hundred times more sky than any pre-vious mission.[11]

TESS packs a lot of scientific punch for its size. Whereas Kepler was the size of a small bus, TESS is the size of a mailbox. Instead of a telescope, it has four cameras with wide-angle lenses looking out in slightly different directions. The open aperture is equivalent to a telescope of only ten centimeters in diameter. Most amateur astron-omers use larger telescopes than this.

Exoplanets found by TESS will be one-tenth as far and a hundred times brighter than exoplanets found by Kepler, so follow-up obser-vations will be much easier. TESS will monitor over a half million nearby stars for dips in brightness caused by planetary transits. By the time all the data has been analyzed, TESS should find fourteen thousand exoplanets, including three hundred super-Earths and a few dozen Earth clones. In January 2020, it found its first Earth-size planet in the habitable zone. The system is unusual because two Earth-size planets sandwich a super-Earth.[12] Familiar names are working on the data. Sara Seager is the deputy director of science and David Charbonneau is in the core science team. In her excite-ment about the data from TESS, Seager says, "We're hoping that it will find the very special Goldilocks planets—the ones that could be the right distance from the star, that have rocky surfaces, that are perfect for following up to look at the atmospheres with another tele-scope."[13] The initial two-year mission has been extended through 2022, and if the hardware holds up, TESS could be discovering exo-planets for decades.

TESS and other exoplanet-hunting missions make use of the foun-dation of astronomy: measuring positions on the sky.[14] The oldest sky maps date to the Babylonians three thousand years ago. In the second century BCE, Hipparchus compiled the first stellar catalog, containing 850 stars measured to an accuracy of one degree, twice the angular size of the full Moon. Major advances in astronomy have often been driven by better angular measurement. Examples include astronomer Tycho Brahe's improved positions for Mars that spurred the Copernican Revolution in the sixteenth century, mathematician Giovanni Cassini's first measurement of the scale of our solar system in 1672, and the first accurate measurement of the distance to a star by astronomer Friedrich Bessel in 1838.

For four hundred years since the invention of the telescope, positional accuracy has grown logarithmically, in another echo of Moore's law. The first space mission to measure star positions led to a big advance when its catalog was published in 1997.[15] The Hipparcos catalog contained positions for 120,000 stars with a precision of 0.001 arc seconds, probing star distances out to three hundred light-years. ESA's current mission, Gaia, is another giant leap forward, aiming to map positions of a billion stars with a precision of 0.00002 arc seconds, or 20 micro–arc seconds. That's equivalent to measuring the width of a human hair at a distance of a thousand kilometers! Gaia will make a three-dimensional map of 1 percent of the entire Milky Way galaxy.

What does this have to do with exoplanets? Gaia data will measure motions using seventy separate observations of each star throughout the mission, which operates until 2025. The exquisite precision of measuring star positions opens the possibility of detecting planets by the wobble they induce on their star. As opposed to using spectroscopy and the Doppler effect, Gaia will succeed by watching the actual wobble of the star compared with nearby stars that have no orbiting planets. A Jupiter-like planet orbiting a sun-like star with a five-year period induces a wobble that from fifty light-years away, appears as an angular motion across 190 micro–arc seconds, which will be easily detectable. That same planet could be detected as far away as a thousand light-years.[16] A Neptune-size planet can be detected out to two hundred light-years. Gaia should net tens of thousands of giant planets when its full data set is made available.[17]

Next up for ESA is PLATO. That's an acronym for Planetary Transits and Oscillations of stars, and a nod to the seminal Greek philosopher. PLATO takes the design of TESS and turbocharges it.[18] The spacecraft is the size and weight of a small truck, and looks like a fly's eye, bristling with twenty-six wide-angle cameras. The CCD behind each camera has eighty million pixels, for a total detector real estate of two billion pixels. The price tag is $400 million, and the launch date is sometime in 2026.[19] PLATO will target a million bright stars for transit observations, trying to home in on the most habitable planets in the sample. The instruments will deliver exquisite accuracy, with measurements of radii to 3 percent precision, masses to 10 percent precision, and ages to 10 percent precision. PLATO is expected to deliver hundreds of Earth twins and super-Earths, all suitable for detailed characterization with large, ground-based telescopes.[20]

NASA is also proposing an ambitious new telescope for probable launch in 2025. The Nancy Grace Roman Space Telescope has a 2.4-meter mirror—the same size as the Hubble Space Telescope—but it will achieve similarly sharp images in the infrared over a one hundred times larger field of view.[21] Originally called Wide Field Infrared Survey Telescope (WFIRST), the Nancy Grace Roman Space Telescope is part of a sequence of space missions by NASA and ESA that will propel the field of exoplanets forward.

The project has a long and intriguing history. In 2011, NASA was pursuing a wide-field space telescope to study dark energy, with a 1.3-meter mirror and single instrument. Former astronaut John Grunsfeld had just started a new job as the NASA associate administrator when he got an unusual phone call. It was a senior official at the agency that runs the nation's spy satellites, asking if he was interested in some free, surplus hardware. The items were twin telescopes the same size as the fabled Hubble Space Telescope, but built for looking down at Earth rather than up at the stars. Grunsfeld

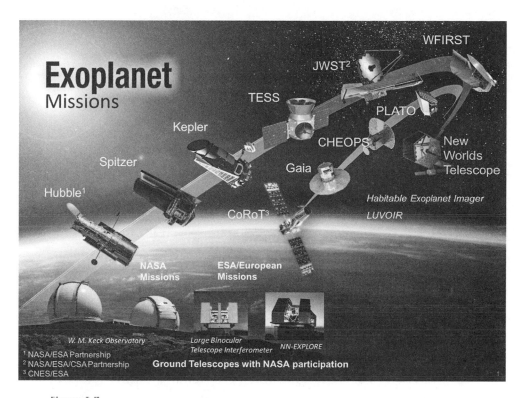

Figure 6.2

Exoplanet missions of NASA and ESA. Ground-based facilities using NASA funding are also shown. WFIRST was renamed the Nancy Roman Space Telescope. (Credit: ESA, "Exoplanet Mission Timeline," last updated 2020, https://sci.esa.int/web/exoplanets /-/60649-exoplanet-mission-timeline.)

initially thought the gifts might be a distraction—hardware expensive to handle and store. But the scientists involved in the 1.3-meter space telescope project were enthused. They realized the optics were exquisite and exactly suited to their goals, having twice the collecting area of their planned mirror. Their project was stalled due to the cost overruns of the JWST, dubbed "the telescope that ate astronomy."[22] This free mirror, however, cut their costs by $250 million. When Grunsfeld asked them if it could be used to study dark energy, they told him it was perfect. Adam Reiss, who won a

Nobel Prize for discovering dark energy, thought it might be too good to be true. Yet he said, "When someone hands you a hand-me-down like that, you have to be excited. They're not sitting around at Walmart."[23]

After many twists and turns, three reincarnations, and two near-death experiences, the Nancy Grace Roman Space Telescope received approval from Congress, with a maximum budget of $3.9 billion and funding for five years of science operations. It's a flagship NASA mission. The imaging camera will use three different types of measurements of a billion galaxies to better understand dark energy, the major ingredient of the universe that is causing cosmic acceleration. It will also survey the inner Milky Way for exoplanets using the microlensing method (see chapter 5). The expected yield of exoplanets that the project will discover by this technique is twenty-six hundred—a huge gain compared to the hundred exoplanets discovered this way to date. Crucially for exoplanet science, the second instrument that was added is a coronagraph (see chapter 4). Its deformable mirror technology is aiming for the factor of a billion-to-one light suppression that will allow imaging of earthlike planets. To Seager's pleasure, it is being built so that a starshade can be added later.

Where does this leave the Earth-bound spectroscopists? After all, they found the first exoplanet and most of the first few hundred, before Kepler discoveries surged. Has all the flavor been chewed out of the terrestrial gum?

Not at all. Transits only measure planet size. To know the mass as well as get a sense of the structure and composition of the planet, Doppler spectroscopy is needed. Also, transits only occur for a small fraction of orientations, when the solar system is nearly edge on, so our sightline is through the plane of the planet orbits. The Doppler method gives a signal for almost all orientations, except when our sightline to the planetary system is perpendicular to the plane of the planet orbits. Spectroscopists like Mayor still have tricks up their sleeves. He notes that "state-of-the-art instrumentation and dense temporal sampling [many and frequent] are the keys to making progress."[24] The goal is a velocity precision good enough to detect Earth-mass planets. That's a stellar wobble of just ten centimeters (four inches) per second, or a leisurely walking speed. How ironic that as planet hunters improve their craft, they home in on quarry moving so slowly, you'd think they could chase them down.

A velocity precision of ten centimeters per second implies an extraordinary wavelength precision of one part in a hundred million. There are two big obstacles to achieving this goal. One is the stability of the spectrograph. Astronomy is not done in a lab under carefully controlled conditions. It's done at a telescope that moves around the sky and is open to the elements. Temperature changes make metal expand and contract, and humidity can interfere with delicate optical elements. Planet hunters have solved these problems by feeding their spectrographs with optical fibers from the telescope and putting them at lower levels of the telescope building in climate-controlled rooms.

The second problem is more fundamental. A planet induces a periodic Doppler shift on the spectrum of a star. But stars are not uniform balls of hot gas. They are churning and variable so the star itself can produce signals that could be interpreted as planets. This behavior is grouped under the term "stellar jitter." It includes convection cells on the star's surface, star spots that come and go, and magnetic activity cycles.[25] These signals are on all time scales from hours to years, so they can interfere with the detection of short-period planets. Astronomers have figured out clever ways to mitigate most of the stellar jitter. Mayor has led the development of increasingly precise spectrographs for thirty years. He's most excited about an instrument that was commissioned on the Very Large Telescope in Chile in 2019.[26] It promises to put the "holy grail" of Earth detection within reach. With exoplanets being discovered in space on an industrial scale, Doppler surveys will have to ramp up dramatically.

To follow up on five thousand targets will take five million observations over ten years and forty expensive, high-resolution spectrographs, most of which haven't yet been built.[27]

The exoplanets will continue to pour in for the next decade. For a moment, let's turn the tables on this outward exploration and pose the question in reverse. If we're looking so hard for habitable planets, is anyone out there looking for us?

Lisa Kaltenegger asked that question. She is the director of Cornell University's Sagan Institute, which she founded when she was only thirty-seven. She and a colleague identified a thousand stars that might host habitable, earthlike planets that all have a special perspective on our solar system. These stars are in parts of the sky where they could detect Earth transiting the Sun.[28] "If observers were out there searching, they would be able to see signs of a biosphere in the atmosphere of our Pale Blue Dot," she said, invoking the phrase Sagan coined for Earth as seen from afar. "If we're looking for intelligent life in the universe that could find us and might want to get in touch, we've just created the star map of where we should look first."[29]

7
Milky Way Census

The progress in discovering exoplanets has been breathtaking. After decades of failure, it took just twenty-eight years to go from zero to five thousand. The sample is already big enough to try to project the population across the entire galaxy. Before we can do that with any confidence, we must understand the limitations that attach to any scientific measurement, and figure out the comparative strengths and weaknesses of the various exoplanet-hunting methods.

An analogy will help explain the difficulties in extrapolating exoplanet discoveries made so far into an estimate of the total number of exoplanets and the relative sizes of the populations of different kinds of exoplanets. It's a fishing analogy.

You're fishing in a small boat on a vast lake. You wonder about the diversity of the fish in the dark depths below you. How many fish are there and what kinds?

For the sake of conservation, let's assume you're practicing catch and release. You have a net and cast it over the side. After a while, you pull it up with a dozen fish in it. One fish is a foot long, a couple are four to six inches long, and most are one to two inches long. You wonder if this means there are many more small fish than large fish in the lake. Then you realize that the net has one-inch holes so it's not able to catch even smaller fish. You can't say anything about the number of tiny fish in the lake. Also, you're fishing in one location. Perhaps your net would gather different types of fish, or different proportions of small and large fish, if you were on the other side of the lake.

Next you try a pole and line. Luckily, you're patient because after an hour, you've caught nothing. Then you catch five fish in close succession. They're all quite large and the same kind, which is not the same kind as any of the fish caught by the net. Questions bubble up. Does this mean your lure can only attract one kind of fish? Would the result be different if you used a worm instead of an artificial lure? Might there be fish too small to take the bait and fish so large that they would ignore it? If you'd caught nothing after three hours, could you have reasonably concluded there were no fish in the lake?[1]

The next day you come back and apply technology to the problem. You have a sonar that you can hang off the side of the boat. Surely this will give you an unbiased and complete census of fish in the lake. You quickly realize it's not that simple. The sonar doesn't give you a nice image of the fish, like a photograph; it gives you a squiggle on the screen that's hard to decipher. The manual says it has a range of fifty meters, but is that the range for a particular size of object, and if so, what size? You realize that the sensitivity must vary with the distance through the water. That evening a marine biologist friend tells you sonar can only receive a signal from the swim bladder of a fish, a gas-filled organ they use to control their buoyancy.[2] Your friend says that fish with skeletons that are mostly made of cartilage, like sharks, don't have swim bladders. Your sonar is not going to spot any sharks that might be in the lake. Suddenly you get nervous about the evening swim you'd been planning.

This extended analogy makes the point that you can only find what your tools will let you find. The net can't catch fish smaller than one inch. The sonar can't see fish more than fifty meters away. In astronomy, if you understand the limitations of the measuring device, you can predict what it will and won't detect, and even make corrections for the objects that are missing from your sample. Not knowing something is not a brick wall! You can make a model of your ignorance, and project what you might have missed.

Although it's even more fanciful, let's compare exoplanets to snarks. In his nonsense poem, *The Hunting of the Snark*, writer Lewis Carroll notes that they cannot be captured in a commonplace way and courage is required to hunt them.[3] The character who leads the hunt tells the crew five signs by which snarks can be identified. Exoplanets make for elusive prey as well. Astronomers use five methods to snare them, and all are needed to understand the true population. Each has strengths and limitations.

The radial velocity (Doppler wobble) method was used to discover the first hundred exoplanets and has found nearly a thousand so far (see chapter 2).[4] This method measures an exoplanet's mass, subject to the unknown inclination of the orbit. The repeating signal gives the orbital period, and Kepler's laws then give the distance of the planet from the star. It is most efficient at finding massive planets in close orbits, and the technique is well suited to ground-based telescopes.[5] Planets are faint, but nearby stars are bright, so the Doppler method can be used with small telescopes. The method is inefficient at finding planets far from their stars, due to the long orbital times, and it can't measure the size of a planet. It fails if we're looking down on the plane of the orbit because then there's no radial component to the motion. Only if our view is in the plane of the orbit will we see the full motion.[6] Since planetary systems are randomly oriented in space, mass is underestimated on average by a factor of two. The Doppler technique runs into a fundamental sensitivity limit around the mass of Earth.

The transit photometry method (chasing shadows) was responsible for the surge in exoplanet numbers in the past decade thanks to the Kepler spacecraft. The method measures planet size and only works if the inclination of the orbit is edge on, where our sightline passes through the plane of the planet orbits. If it's even slightly tilted, the planet passes above or below the star, and there's no eclipse. In fact, for over 95 percent of the random orientations of a planetary system, there's no eclipse. So how was Kepler so successful?

It crushed the problem with numbers. By staring at 150,000 sunlike stars at once, the small fraction with planets in a suitable orientation was still a hefty number. Transits work well for finding smaller planets. About 85 percent of Kepler's discoveries are smaller than Neptune, while 85 percent of the exoplanets found by other methods are larger than Neptune.[7] Transits don't measure planet size directly. The percentage dip in the starlight gives the ratio of planet cross-sectional area to star cross-sectional area. You need to know the star's size to deduce the planet's size. Which immediately raises a problem. Stars are so far away that we can't measure their sizes directly. Proxima Centauri, 4.4 light-years away and the nearest star to the Sun, has an angular size of 0.0001 arc seconds.[8] As we've seen, that's far too small to measure with a telescope even if it's above Earth's atmosphere. Most Kepler targets are hundreds of light-years away, so their angles would be millionths of an arc second. Astronomers use spectra of the stars to figure out their mass and evolutionary state, and then they compare to models to get the size.

The transit method is most efficient at finding planets in close orbits, and the technique requires the stability afforded by observing from space. It's ineffective at finding planets at most orbital inclinations and can't measure a planet's mass. That's a significant limitation, as we can see if we look at the relationship between planet mass and size. Rocky planets are predictable. As the size grows, the volume grows by the size cubed, and the mass grows the same way. In other words, one planet twice the size of another will have eight (two cubed) times the volume and therefore eight times the mass. But gas giants are mostly made of gas, and gas compresses as more is piled on, so they grow more slowly as they get more massive. For example, in our solar system, Jupiter is three times Saturn's mass but only 20 percent larger in volume. For planets above Jupiter's mass, the compression is so strong that they get *smaller* as they get more massive.[9] So transits are a poor indicator of the mass of giant planets. Worse, the highest mass a planet can have is heavier than the lowest mass a star can have. An enormous Jupiter-like planet can be more massive than a small "failed star" not hot enough to shine by fusion (a brown dwarf). As a result, there's a twilight zone where it's ambiguous whether a transit dimming is caused by a planet or star.[10] Instead of chasing shadows, we're chasing our tail. At least for small, terrestrial planets, size measurements from transits are "rock solid."

Direct imaging is the most obvious search method, but the hardest in practice due to the enormous brightness ratio between a star and its nearby planet. Imaging is direct evidence, and the orbit can be measured with a sequence of images. Neither mass nor size, however, can be derived from these observations. The method is most effective at finding planets in large orbits and with inclinations that are face on, so our sightline is perpendicular to the plane of the planet orbits and the planet doesn't cross the star. It's not as effective for planets that orbit luminous stars. It also gets challenging for more distant targets, where the angular separation between the planet and star becomes unmeasurably small. Fewer than 3 percent of exoplanets have ever been imaged, and astronomers are putting huge efforts into doing better with clever optics (including chronographs and, perhaps one day, a starshade).

Gravitational lensing is the method that makes use of the brightening of one star when another star passes in front of it. It works thanks to the bending of light, as described by general relativity (see chapter 5). If the nearer star has a planet, another amplification signature is added (and an exomoon would add another, yet smaller signal). Unlike all the other methods, lensing is a onetime event. The method works well for distant planets, planets far from their stars, and even free-floating planets that have somehow gotten detached from their stars. It's not efficient for finding many exoplanets at the same time.

The final method is astrometry—measuring the positions of stars with extreme precision and following how they move. Finding exoplanets using astrometry involves observing the way a planet tugs at its star. Rather than looking for a periodic Doppler shift in the star's spectrum, we look for the actual wobble of the star in the sky. Astrometry can show a star moving irregularly compared to nearby stars. This wobble gives an estimate of the planet's mass. The method works best for detecting planets with large orbits and those that don't cross in front of the star. It's not effective for planets at large distances from us and cannot measure diameters. Unlike the radial velocity and transit methods, astrometry isn't well suited to finding multiple planets in one system.

I've left astrometry until last because it's brutally difficult. Recall that Jupiter makes a star like the Sun pivot around its edge, so the wobble would be the same as the angular size of a star like the Sun. That's less than 0.001 arc seconds for even the nearest star. Attempts to see the wobble were star-crossed for many years. In the late eighteenth century, astronomer William Herschel claimed that an unseen companion was influencing the motion of the bright star 70 Ophiuchi. It took a century for this claim to be refuted.[11] In 2002, astrometry with the Hubble Space Telescope was used to characterize a previously discovered planet around the star Gliese 876.[12] Finally, in 2013 the first exoplanet was claimed using this method.[13] Yet at thirty times the mass of Jupiter, it's almost certainly a brown dwarf and not a planet. We'll have to wait for the Gaia satellite to open the floodgates on detecting planets with astrometry.

Having reviewed the methods by which astronomers are so avidly fishing for worlds outside our solar system, we're in a better position to consider the number and diversity of the overall population within the "lake": our own Milky Way galaxy. One-third of all the exoplanets found so far are closer than a thousand light-years and 90 percent are closer than four thousand light-years. In a galaxy a hundred thousand light-years across, all the discoveries have been in our backyard.[14] With over five thousand exoplanets confirmed, and another nine thousand likely to be confirmed, the statistics are sturdy enough to extrapolate from the neighborhood to the entire Milky Way.

The goal is to convert these numbers—a body count of exoplanets of different types—into a fraction of all the stars in the galaxy that have that type of exoplanet. That means understanding the limitations of the surveys—what kinds of fish get through the holes, which ones evade the sonar, and so on. Kepler has generated most of the discoveries, so its data is central to this analysis. By 2013, an early analysis showed the relative abundance of planets of different sizes.[15] One in six stars has an Earth-size planet on a tight orbit, one in four has a super-Earth on a wider orbit, one in four has a mini-Neptune on an orbit up to 250 days long, and only one in twelve has a gas giant planet on an orbit of 400 days or less.[16] The surprise in these statistics is the abundance of super-Earths and mini-Neptunes—the most abundant planet types detected in the galaxy, but ones that don't exist in our solar system.

The broadest net for exoplanets is cast by the microlensing method. Within a broad range of orbital distance from the host star—from 0.5 to 10 AUs (recall that 1 AU is the Earth-Sun distance)—a microlensing survey found Jupiter-like planets around one in six stars, and more than half of all stars hosting a super-Earth and mini-Neptune.

The conclusion was that every star in the Milky Way has at least one planet.[17] In other words, there are more planets than stars!

Recently, researchers have squeezed all the goodness out of all ten years of Kepler data. They've counted the planets that are in the habitable zones of their stars, at a distance where liquid water can exist. That's the Goldilocks zone where the conditions are just right for biology. About 60 percent of sunlike stars have a habitable, earthlike planet—and given the uncertainties, the percentage may be as high as 90 percent. That implies there should be roughly four habitable planets nearby, within thirty light-years of Earth.[18] Another study looked at cool dwarf stars, which are much more abundant than sunlike stars, and found that a similar fraction has

Known Transiting Planets by Size

As of May 10, 2016

PLANET SIZES OBSERVED IN OUR SOLAR SYSTEM

MERCURY • • MARS   VENUS •• EARTH   NEPTUNE   URANUS   SATURN   JUPITER

■ Newly validated Kepler planets
■ Previously verified planets

Number of Planets

1000 — 800 — 600 — 400 — 200 — 0

Mars-size (.5 - .7 R⊕)
Earth-size (.7 - 1.2 R⊕)
super-Earth-size (1.2 - 1.9 R⊕)
sub-Neptune-size (1.9 - 3.1 R⊕)
Neptune-size (3.1 - 5.1 R⊕)
sub-Jupiter-size (5.1 - 8.3 R⊕)
Jupiter-size (8.3 - 13.7 R⊕)
super-Jupiter-size (13.7 - 22 R⊕)

habitable planets.[19] Statistical projections like these still have to be confirmed by actual discoveries.

This is a stunning conclusion. With 400 billion stars in the galaxy, there are more than 400 billion planets altogether. Specifically, and more pertinently to the issue of life in the universe, there are probably 4 billion earthlike planets in situations like ours in the Milky Way, and 20 billion more if we include Earths orbiting the larger number of red dwarfs. Even these numbers are almost certainly an undercount since Earths have only been found around single stars and half of all stars are in binary systems, in which the second star makes planets harder to detect.[20] Now recall that there are 100 billion galaxies in the universe, and ours is typical. Projecting beyond the Milky Way, there are roughly 1,000 billion billion—that's a 1 with 21 zeros after it (10^{21})—potential biological experiments in the universe.

These numbers sunk in for me during a vacation in Hawaii. Lying under the shade of palm trees on Waimanalo Beach in Oahu, the arc of pristine white sand stretched for miles in each direction. My mind was lazy and wandering, but it snapped to attention as I considered the vast number of grains of sand on any beach. I did some quick mental math. My beach towel covered about a square meter. Assuming the sand below it was one meter deep, that one cubic meter held fifty billion sand grains, about the number of habitable worlds in the Milky Way. I imagined how long it would take to inspect each sand grain for any signs of life. More mental math. The sum of all the sand grains on the entire arc of the beach was a number similar to the number of habitable worlds in the universe. I

Figure 7.1
The size distribution of exoplanet discovered by the Kepler satellite. The two most common types of exoplanets detected in the nearby Milky Way, super-Earths and sub-Neptunes, are absent in our solar system. (Credit: NASA Ames and W. Stenzel, in "NASA's Kepler Mission Announces Largest Batch of Planet Discoveries Ever," last updated 2016, NASA Ames, https://exoplanets.nasa.gov/news/1346/nasas-kepler -mission-announces-largest-batch-of-planet-discoveries-ever/.)

mulled this fantastically large number as waves rhythmically lapped on the beach. And I recalled the words of my countryman, Scottish philosopher Thomas Carlyle: "If they be inhabited, what a scope for misery and folly; if they be uninhabited, what a waste of space!"[21]

Considering the mind-blowing figures above, surely we're not alone. There must be civilizations that have been down roads like our own and confronted the challenges we're facing—ones that have struggled as we have (and still do) to survive and thrive. The exoplanet census motivates the search for extraterrestrial intelligence. With billions of Earth clones in our galaxy, many of which formed billions of years before Earth, it would have to be an incredible fluke for Earth to be the first or only planet with biology. And if biology is abundant in the galaxy, evolution to intelligence and technology would have to be a unique fluke if there's nobody out there to talk to.

II
Habitability and the Exoplanet Zoo

To gain perspective on the home planet, we explore the diversity of the many worlds discovered beyond our solar system. The solar system is our familiar reference point, but some of these new worlds are unlike any that orbit the Sun. Their life stories—how they formed and how they have evolved over cosmic time—is a context for a better understanding of the unique properties of Earth. I have described the discovery phase, in which exoplanets are characterized by a single attribute: size or mass. Now we learn more about their properties and view them through the lens of habitability: their potential to host biology.

Life on Earth began when this planet was a hot cauldron of volcanoes and impacts from space. Over four billion years, life has radiated into every conceivable evolutionary niche, filling the full envelope of physical conditions on land and sea. We don't know if life can emerge in the bizarre environments that astronomers have discovered beyond our solar system. Earth is the place where we evolved, but it's not the "best of all possible worlds." These chapters tour the exotic inhabitants of the exoplanet "zoo," where technology will soon be good enough to characterize these alien worlds. The questions I want to answer: Could they be inhabited, and if so, by what?

8

Gas Giants

The first exoplanet to be discovered—a planet orbiting 51 Pegasi—was a gas giant 50 percent larger than Jupiter and closer to its host star than Mercury is to our Sun. It's hot enough to melt aluminum, and it moves around its star on an insanely fast four-day orbit. There's nothing like this "large, hot Jupiter" in our solar system, so the first exoplanet was as perplexing as it was exciting. The second and third exoplanets ever found were also hot Jupiters—and so were ten of the first twelve.

It is not surprising that the planet hunters would find large planets on tight orbits first since they induce larger Doppler shifts on their host stars. As more discoveries came in, using the other methods described in the preceding chapters, we gained a clearer census of all the fish in the lake. It turns out that hot Jupiters are quite rare. Only 1 percent of all stars have one. The real question is why there are any at all.

There's a soothing regularity to the orbits of planets. The planets that are visible with the naked eye—Mercury, Venus, Mars, Jupiter, and Saturn—wheel through the sky on cycles that never change. They seem eternal and imperturbable. Shakespeare knew this well:

> The heavens themselves, the planets, and this center
> Observe degree, priority, and place,
> Insisture, course, proportion, season, form,
> Office, and custom, in all line of order.[1]

We have a story of how our solar system formed that makes physical sense. About 4.5 billion years ago, a huge, nebulous cloud of gas and dust collapsed, possibly caused by the shock wave from a nearby exploding star. As the cloud shrank by gravity, the small amount of rotation was amplified, like the spinning of an ice skater when they bring their arms closer to their body. This spinning, swirling disk was the solar nebula. At its center, gravity created heat and pressure so great that atoms fused to form helium. The Sun was born, and it eventually pulled in 99 percent of the available matter.[2] Material farther out was clumping and steadily forming larger and larger objects. The gravity of the biggest ones was sufficient to make them spherical, so they became planets and large moons. Occasionally no planet formed; the asteroid belt contains insufficient mass to make a planet. Other leftover material became comets, meteors, and small moons. These objects are made mostly of rock, mixed with water ice and frozen gases.

The pleasing regularity of our solar system is due to the way it formed. Planets spin in the same sense that they orbit the Sun.[3] Miniature versions of the collapse process caused planets to have moons orbiting them. Nearest to the Sun, only rocky material could withstand the heat when our solar system was young. For this reason, Mercury, Venus, Earth, and Mars are terrestrial planets, small with rocky surfaces. Farther out, there were also rocky planets, but the cooler conditions meant the rocky cores were mixed with ice and frozen gases, and the larger amount of material available allowed gravity to add large envelopes of gas to the rocky cores. Far from the Sun, we find the gas giants, Jupiter and Saturn, and the ice giants, Uranus and Neptune.

This formation scenario is called the nebular hypothesis. The idea started with Emanuel Swedenborg, who was born in Sweden in the same century that Shakespeare died. Swedenborg was a prolific inventor and scientist, coming up with designs for a flying machine and a concept for neurons.[4] In midlife, he entered a spiritual phase where he experienced visions and claimed many psychic experiences. German philosopher Immanuel Kant was intrigued by Swedenborg's reputation as a mystic. Although he later was scathing about Swedenborg's purported psychic powers, Kant was intrigued by the nebular hypothesis and developed it further.[5] The idea had successes, but also some problems, as it was refined over the next two hundred years. It took its modern form thanks to Russian scientist Victor Safronov.

With a brilliant mathematical treatment, Safronov showed how the process of accretion could turn dust bunnies into rocks, then mountains, and then planets. He worked in almost complete isolation in Russia during the Cold War. In the early 1970s, work he had done decades earlier came to the West by a circuitous route and was translated into English, becoming the basis of how astronomers think planets form.[6] But there was still a lingering malaise in the scientific community. What if this clever hypothesis was a "just so" story: a convenient narrative engineered to fit what we see around us?[7] The idea could only be truly tested on another planetary system.

That's why it was such a surprise that the first exoplanet discovered was a large, hot Jupiter, orbiting so close to its star. It blew the tidy narrative of planet formation out of the water. In the nebular hypothesis, planets form by sweeping up material in "zones" of a disk of gas and dust. Close to the star, there's only enough stuff to make small rocky planets. Farther out, those rocky planets grow to several times Earth's mass and then they attract enough gas so that most of their final mass is gas, not rock.[8] Around the time the nebular hypothesis was first posed, astronomers noticed that the spacing of planets is geometric; each planet is approximately twice as far from the Sun as the one before. It was another example of Shakespearean proportion in the motions of the heavens. The spacing of the planets shows why hot Jupiters are so surprising. The surface area of material that can be swept up to make a planet grows rapidly moving outward in our solar system. The zone at Jupiter's distance is over two hundred times larger than the zone at Mercury's distance. It's hard to see how a planet with the heft of Jupiter could grow at the location of Mercury. Even if it did, intense solar radiation would blow away most of its atmosphere and limit its size.[9]

If they didn't form at their current locations, hot Jupiters must have migrated in from farther out. The physics of migration is tricky; turning a cold, far-out Jupiter into a hot, close-to-the-Sun Jupiter means reducing its angular momentum by a factor of ten and its orbital energy by a factor of a hundred.[10] Migration only works if it happens soon after the planetary system forms when there's still gas left in a disk. The planet grows a rocky core by the process of accretion, and then gathers a mantle of hydrogen and helium to become a gas giant. The process is rapid, taking less than 10 million years, which is the blink of an eye in the 4.5-billion-year history of our solar system.

Hot Jupiter exoplanets are puffier than our familiar Jupiter. They have the density of Styrofoam, which means if you put one in a large enough bathtub, it would float. After it forms, it starts to feel a drag as it moves through the gassy disk. Imagine a large person struggling to move through a thick liquid. The interaction of the giant planet with the disk makes it lose energy and angular momentum, and it spirals in toward the star. Migration is a struggle. The giant planet can lose half of its mass plowing through the disk material, and on the way in it shoves aside and ejects planetary embryos and chunks that can form terrestrial planets. An interesting question is whether a solar system with a hot Jupiter can also have an earthlike planet at an earthlike location? The answer is yes. Although the migrating giant will scatter material like tenpins, the disk can reform, and there's plenty of stuff left and plenty of time afterward to grow an Earth.

Somewhere out in space, there might be creatures on an Earth clone with no doubt that planets can migrate because it happened in their backyard.

Migration leads to a new puzzle: If Jupiters can migrate close to their stars, why don't they just keep spiraling in and disappear? Jupiter is one-thousandth of the Sun's mass, so consuming a Jupiter is just like eating snack food for a star such as the Sun. The answer seems to be that when the planet gets close in, tidal gravitational forces make the orbit circular and lock the planet in a death grip with the same side facing the star. The strong stellar radiation drives away gas from the planet's atmosphere. Intense gravity pulls the planet into the shape of a rugby ball. A milder form of this tidal locking keeps Mercury with one side always pointing toward the Sun and keeps the Moon with one side always pointing toward Earth.[11] The most extreme member of this club is K2–137b. It has an orbit twenty times faster than the prototypical hot Jupiter, 51 Peg b. It whirls around its star every four hours at a distance of half a million miles. Astronomers have deduced that iron accounts for nearly half of its mass, which means the star must have blasted away its normal gas giant atmosphere. Imagine a giant ball bearing 150 times the mass of Earth hurtling through space at nearly a million miles per hour![12]

The discovery of hot Jupiters was a kick in the pants to theorists and a reminder that they've often scrambled to explain what the planet hunters are finding. As in other areas of astronomy, observations led the way. Several theorists had an inkling that planets could migrate, but they didn't make the prediction. One of the leading planet theorists, Alan Boss of the Carnegie Institution in Washington, DC, was asked if he ever pinched himself for not predicting planet migration. He replied, "Oh, my body is heavily bruised from all the pinching, and everyone else as well. Observers generally love to gloat about victories over theorists. They have a lot to gloat about. Paul Butler likes to say that not a single theoretical prediction has been borne out since the discoveries of exoplanets. That's true."[13] (Butler was the discoverer of the second-known exoplanet; see chapter 2.)

Discovery methods for exoplanets yield limited information. Doppler spectroscopy gives the mass, and a transit indicates the size. Combine the two, and you get the average density—a calculation that has been done for several hundred exoplanets. A single average density isn't good enough to characterize the structure of a planet with different layers, but it's good enough to say whether a planet is dominated by rocks, like Earth, or gas, like Jupiter. To characterize exoplanets in more detail, astronomers turn to spectroscopy.[14] When a planet transits a star, the light dims slightly and briefly. There's another effect, however. Backlit by the star, the rocky mass of a planet blocks the star's light, but some starlight filters through the atmosphere and reaches us. As it passes through the planet's atmosphere, the starlight is imprinted with spectral features that reveal the chemical composition of the planetary atmosphere. Yet starlight also has spectral features. To tease out information about the planet, one spectrum is taken when the planet is behind the star, and another when it's in front of the star. The "extra" absorption seen when the planet is in front of the star is caused by the atmosphere of the planet. This clever method has been used to learn what exoplanet atmospheres are made of for the first time.[15]

The first exoplanet to have its atmosphere "sniffed" in this way was HD 209458b, a hot Jupiter that was also the first exoplanet to have a transit measured.[16] The dominant gases in hot Jupiter atmospheres are the elements hydrogen and helium, but interesting trace gases are found. Water was the first molecule to be detected. While much hotter than any sauna, these atmospheres are like a sauna in that droplets called aerosols condense within the hot gas. There's steam on hot Jupiters. Beyond that, this air is exotic and not like anything we might encounter on Earth. The most common ingredients are silicates—tiny particles of sand—along with hydrocarbons like ethane and methane. There are also metal oxides made with chromium, titanium, and iron. Toss in a pinch of sulfur and soot, and we get a whiff of Hades.[17]

One place to avoid is HD 189733b, the first exoplanet to have its atmosphere mapped. Its weather is deadly. Howling winds of nine thousand kilometers per hour (fifty-four thousand mph) would pelt a visitor with molten glass.[18] On the hottest of the hot Jupiters, metal would vaporize. Consider WASP-76b. On the scorching day side of this planet, the temperature is 2,400°C (4,350°F), and

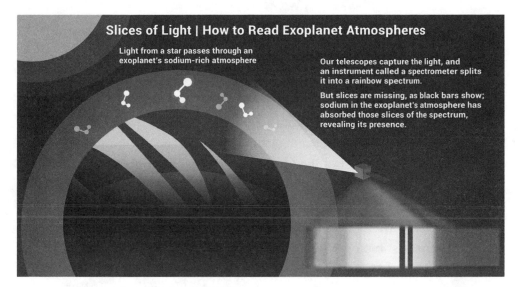

Slices of Light | How to Read Exoplanet Atmospheres

Light from a star passes through an exoplanet's sodium-rich atmosphere

Our telescopes capture the light, and an instrument called a spectrometer splits it into a rainbow spectrum.

But slices are missing, as black bars show; sodium in the exoplanet's atmosphere has absorbed those slices of the spectrum, revealing its presence.

Figure 8.1
Spectroscopy can reveal the chemical composition of an exoplanet atmosphere. The exoplanet is backlit by its star, and absorption in the spectrum occurs at wavelengths specific to particular elements. (Credit: NASA/JPL-Caltech/L. B. De La Torre, "Spectroscopy Infographic," last updated 2021, https://exoplanets.nasa.gov/resources/2270/spectroscopy-infographic/.)

iron exists as a vapor. Fierce winds carry the metal to the night side, where it falls as a rain of molten iron droplets.[19] KELT-9b is even hotter. It whips around its star every thirty-six hours and has a temperature of 4,300°C (7,800°F). That's hotter than the surface of many stars and enough to boil any metal, including titanium. These planets cannot last forever; they're being steadily vaporized.

Astronomers have detected cold gas giants too. But these gas giants are undercounted for several reasons. Gas giants would normally be far out in a solar system, requiring decades of orbital data to detect them by the Doppler method. Their odds of making a transit are miniscule, so that method is inefficient at finding them. Yet nearly a half century of data has finally revealed Jupiters around other stars that resemble our familiar Jupiter. NASA keeps statistics on the different types of exoplanets.[20] As of mid-2022, roughly equal thirds of the five thousand confirmed and nine thousand candidate exoplanets were gas giants like Jupiter, ice giants like Neptune, and super-Earths—with smaller terrestrial planets such as Earth accounting for a small minority at around 4 percent.

This is the raw body count; the true percentages in the population will change as we account for all the fish in the cosmic lake using all available detection methods. The field is still young; planet hunters and theorists alike would never rule out more surprises!

This ends my tour of the first category of exoplanets. If the goal was finding habitable locations, the pickings here are slim. Gas giants are either boiling or freezing. It's extremely unlikely that any of them host biology of any kind.

Or is it? In 1976, along with astrophysicist Ed Salpeter, Sagan wrote a provocative paper about life in the Jovian atmosphere.[21] They began by noting that Jupiter has a temperate layer between the frigid cloud tops and the suffocatingly hot and dense base of the atmosphere. Knowing the atmosphere was rich in organic molecules, they envisaged an ecological niche for biology in convection zones of the midlevel atmosphere. They described three types of Jovian organisms. Sinkers were constantly falling, but they lived long enough to produce offspring that would stay in the more habitable layers to repeat the cycle of life. Floaters were several kilometers across, larger than whales and indeed almost the size of cities. Drifting in herds across the alien sky, they would look like immense balloons. Hunters were fast and maneuverable, preying on floaters for their organic molecules and stores of pure hydrogen. The paper was fanciful yet rigorous, and it appeared in a prestigious journal. Since that paper, we've learned about microbial life in Earth's stratosphere, which survives in a similarly rarified and extreme environment.[22] Sagan's ability to mix whimsy and hard science was unparalleled. His biographer was impressed: "They produced one of the more singular articles of the time, a quantitative analysis (with sixty equations) of life, love, and death in the air of Jupiter."[23]

9

Ice Giants

One step down from the gas behemoths are the ice giants. While the mythical prototype for gas giants is Jupiter—king of the gods and chief deity of the largest empire on Earth, under the Romans—the prototype for ice giants is Neptune, the brother of Jupiter and god of the sea.

To Jupiter, consuls swore their oath of office, and generals made ritual sacrifices of oxen, lamb, and goats for success in battle. Meanwhile, Romans celebrated Neptune's festival at the height of summer so as to sustain them in times of heat and drought. Neptune was supposedly a vengeful and vindictive god, with a powerful and stormy temperament like that of the sea. He tried to overthrow his brother but failed. In ancient Rome, three temples were built to honor Jupiter and only one to honor Neptune.[1]

Our solar system is home to two ice giants. But how common are these large, ice-cold planets in planetary systems outside our own?

Planets seem lonely and aloof. The distances between them are vast compared to their sizes. Our solar system is mostly empty space. But the gravitational interactions between planets are subtle and profound. Shakespeare described this as the music of the spheres:

> Sit, Jessica. Look how the floor of heaven
> Is thick inlaid with patines of bright gold:
> There's not the smallest orb which thou behold'st
> But in his motion like an angel sings,
> Still quiring to the young-eyed cherubins;
> Such harmony is in immortal souls;
> But whilst this muddy vesture of decay
> Doth grossly close it in, we cannot hear it.[2]

In these lines, the Bard is referring to an idea from Greek mathematician and philosopher Pythagoras—and it's a notion that has resonance with our modern understanding of our solar system. According to legend, Pythagoras was intrigued by the harmonious ringing of an anvil in a blacksmith's shop. Each hammer produced a unique pitch depending on its weight. Pythagoras was led to the concept of musical intervals, where the difference in pitch between two notes can be reduced to mathematical ratios. Music emerges from the vibrations and oscillations of mechanical objects—the strings of a guitar, wooden bars of a xylophone, or hollow cavity of a flute. Pythagoras believed that mathematics was the foundation of reality, so he extended this idea into the heavens. The Sun, Moon, and planets each produced a distinctive note; together they created the "harmony of the spheres."[3]

The idea was taken up again in the seventeenth century by Kepler. Science and the arts were not segregated then as they are now, and for centuries mathematics, music, and astronomy were core subjects in any university curriculum. Kepler spent years trying to describe the motions of the six-known planets with musical intervals and harmonies.[4] To him, this was more than a mathematical exercise; it had an aesthetic and spiritual dimension. He published his work in the same year as Shakespeare's words appeared in the second quarto of *The Merchant of Venice*. To his great excitement, he converted the varying velocity of each planet into a musical pitch, and discovered the mathematical relationship between the period of an orbit and distance from the Sun—his third law of planetary motion.[5] As we'll learn, he only scratched the surface of the harmonies of our solar system.

To see the differences between a gas giant and ice giant, we can look to our solar system. Jupiter, Saturn, Uranus, and Neptune appear similar; the distinctions lie within. Jupiter and Saturn are 90 percent hydrogen and helium, with rocky cores roughly three to five times Earth's mass. At the base of Jupiter's atmosphere, the pressure is so high that the hydrogen is squeezed into a dense fluid. On Earth, hydrogen is a colorless, transparent gas. In Jupiter, it turns into a bizarre liquid that acts like a metal.[6] A mere 20 percent of Uranus and Neptune are hydrogen and helium. In addition to rocky cores, hidden under the clouds are substantial mantles made of water, ammonia, and methane. Under pressure, these molecules are liquids, but these planets are called ice giants because the ingredients were frozen when the planets formed. Two-thirds of their mass is in the form of vast water oceans. Much of the water is in a strange state, under so much pressure that gas and liquid coexist. Water molecules in this situation separate into hydrogen and oxygen ions, and this form of "superionic ice" conducts electricity well and endures high temperatures before melting. Researchers have recently re-created superionic ice, at twenty-five thousand times Earth's normal sea level atmospheric pressure, using a diamond anvil.[7] To set the scale of ice giant planets, Neptune is four times Earth's size and seventeen times Earth's mass.

If exoplanets can migrate to become hot Jupiters, maybe the gas giants in our solar system moved too? And behind this is a big question: How normal is our solar system?[8]

The solar system is not like "clockwork," with rigid cogs set in place, unchanging. Planets don't remain in their same orbits forever. As planets interact by their mutual gravity, the state of the solar system can change over time. Mostly, the changes are small and slow, but when there are close encounters or resonances, the effects can be dramatic. An orbital resonance occurs when two orbital periods are related by the ratio of small integers.[9] The best way to understand an orbital resonance is to think of pushing a child on a swing. If you push the swing at random times, sometimes you'll boost the motion, and sometimes you'll suppress it, so the pushes average out and the motion isn't strongly affected. But if you push the swing every time at the same point in its motion, or every two times or three times, the motion is increased.[10]

Planetary scientists always had sneaking suspicions that the major planets were not always where we see them now. These suspicions were developed and codified in the Nice model in 2005.[11] "Nice model" was coined not because the idea is pleasant but instead named after the city in southern France where a research group developed it. The model begins with Jupiter, Saturn, Uranus, and Neptune more tightly spaced and closer to the Sun than they are now. Beyond those giant planets are many small bodies of rock and ice, the same asteroid-size pieces from which the planets were built. Some of the thousands of icy rocks wander inward. When they encounter Neptune, they head more strongly inward and push Neptune out a bit. When they get to Uranus, they scatter inward even more, moving Uranus out a bit. The same thing happens as the icy rocks reach Saturn.[12] Each of these encounters has a small effect, but after thousands of encounters, the gas giant planets have moved substantially outward. Jupiter, however, is massive and obdurate, too heavy to be easily pushed around. It's more likely to fling icy rocks far out into our solar system, while it migrates in a bit. So Jupiter moves in while the other three giant planets move out.

What happens next is swift and dramatic. As Jupiter and Saturn reach a state where Jupiter completes two orbits while Saturn completes one, they enter a resonance. In this situation, their gravitational push on each other increases. Jupiter is the bully. It shoves Saturn to its current position, and then Saturn shoves Uranus and Neptune to their current positions. Their outward migration also sends some of the icy rocks farther out, where we see them today as the Kuiper Belt and Oort cloud, home to and source of comets. Others are sent into the inner solar system, where they cause a spike in the bombardment on the inner planets, which could have severely affected the habitability of Earth. Additional resonances explain why Mars is such a puny planet, and why the asteroid belt holds a lot of large rocks rather than a proper planet.[13]

If this sounds like another just so story, rest assured that it's supported by simulations of these interactions and motions, which reproduce the current layout of our solar system.[14] In simulations of exoplanet formation, resonances are equally important, playing a role in the formation of super-Earths and gas giants.[15]

Planetary motions are not sedate clockwork; they resemble a crazed game of pinball. The music of the spheres permeates our solar system. Dozens of orbital resonances have been identified, spanning planets, moons, asteroids, and ring systems. One of the

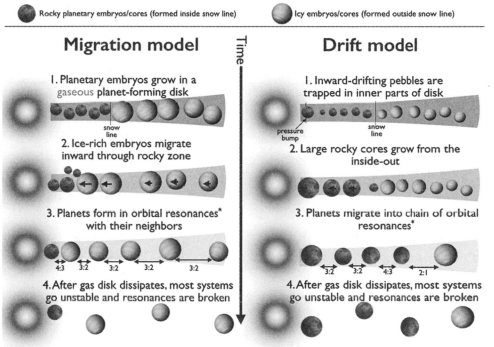

Rocky planetary embryos/cores (formed inside snow line)   Icy embryos/cores (formed outside snow line)

Migration model   Time   **Drift model**

1. Planetary embryos grow in a gaseous planet-forming disk

snow line

2. Ice-rich embryos migrate inward through rocky zone

3. Planets form in orbital resonances* with their neighbors

4:3   3:2   3:2   3:2   3:2

4. After gas disk dissipates, most systems go unstable and resonances are broken

1. Inward-drifting pebbles are trapped in inner parts of disk

pressure bump   snow line

2. Large rocky cores grow from the inside-out

3. Planets migrate into chain of orbital resonances*

3:2   3:2   4:3   2:1

4. After gas disk dissipates, most systems go unstable and resonances are broken

*Resonance (X:Y) — An alignment in which the inner planet completes X orbits for every Y orbits of the outer planet

Figure 9.1
Exoplanet formation scenarios from computer simulations. Super-Earths form in resonance with their neighbors. Giant planets grow, migrate inward, and then are forced outward due to orbital instabilities. (Credit: S. Raymond, in "Solar System Formation in the Context of Extra-Solar Planets" by S. N. Raymond, A. Izidoro, and A. Morbidelli, last updated 2018, https://arxiv.org/pdf/1812.01033.pdf, 10.)

sweetest harmonies involves three of Jupiter's large moons. Io, Europa, and Ganymede have orbital periods in the ratio 1:2:4.

Astronomers are like naturalists exploring a new continent: they look for familiar species, but are often surprised and confounded by what they find. The exoplanet "zoo" contains many strange species, not just hot Jupiters. As the numbers piled up, over a third of exoplanets were ice giants like Uranus and Neptune. That still leaves us ignorant because Uranus and Neptune are the most poorly understood planets in the solar system. Their pale blue hue is caused by methane in their atmospheres, but they might also contain ammonia and water. We don't know their internal structures, the size of their rocky cores, or why Neptune is warmer than Uranus, despite being farther from the Sun. To learn more about these ice giant prototypes, NASA is preparing for the first mission to Uranus and Neptune since the 1980s. Jupiter, Uranus, and Neptune will experience a rare alignment in the mid-2030s that will let spacecraft engineers use Jupiter as a gravitational "slingshot" to boost the spacecraft speed, and reach Uranus and Neptune in a dozen years.[16]

The abundance of ice giants in the exoplanet zoo means we have two thousand of these strange worlds to understand. Although many of these gassy planets are in frigid regions remote from their stars, the word "ice" as used by planetary scientists is misleading. They use ice to refer to any substance that would be gas close to a star, and liquid or solid farther out. It could include a mixture of water, methane, and ammonia, not just water ice. And these planets are not ice balls with thin atmospheres on top. As noted above, they are probably made mostly of highly conductive superionic ice.

One of the Kepler mission's most surprising discoveries was many mini-Neptunes, two to four times Earth's size and ten times its mass. Some researchers have dubbed them Neptinis. These modest ice giants are the most common exoplanets detected.[17] That's striking because in the solar system, Uranus and Neptune are four times the size of Earth and Venus, with no planet in between. Uranus and

Neptune are also about fifteen times the mass of Earth and Venus, with no planet in between. Once again, the solar system seems to be an oddball.

It's poetic that the spacecraft named after Johannes Kepler found echoes of his harmony of the spheres four hundred years later. Kepler-80 is a cool red dwarf star twelve hundred light-years away in the constellation of Cygnus. In 2012, this was the first star found to have as many as five orbiting planets. Five years later, deep learning and artificial intelligence developed by Google were used to raise the number to six planets.[18] The same method was used to find an eighth planet around Kepler-80, putting it into a tie with the Sun as the star with the most known planets. Five of Kepler-80's planets are locked in resonance. The orbital ratios from inner to outer are 4:6:9:12:18. Another impressive harmony was discovered in 2021. TOI-178 is a star two hundred light-years away in the constellation of Sculptor. Five of its six planets are engaged in a gravitational dance, with orbital periods in the ratio 3:4:6:9:18. They are mini-Neptunes with masses three to eight times that of Earth.[19]

Matt Russo is a musician and astrophysicist who has explored sonification, which is the process of converting astronomy data into sound. He'd heard claims that sonification doesn't produce useful science, so he took on the challenge of creating music from all of Kepler's multiplanet systems. The result is a nine-minute YouTube video.[20] The harmonics stand out readily among other dissonant chords. Astronomers gained physical insights from Russo's work and have used it to identify systems for future study. Russo explains what inspires him to meld astronomy and music: "It would be a great present if we could give this back to Kepler. He was looking for musical patterns in our Solar System. To show him that we've found musical patterns in the form of orbital resonances in other systems—I would like to see his face if he could hear that."[21]

Another surprising result from the Kepler mission was the distribution of exoplanet sizes. In general, there are more small planets than large planets. That's not unexpected; it takes more material to assemble a large planet than a small one. The distribution of sizes rises rapidly below six to eight Earth radii, reflecting the preponderance of mini-Neptunes. But it peaks from two to three Earth radii, and then dips with a factor of two deficit from one and a half to two Earth radii, before rising again. There are two populations of small exoplanets.[22] "This is a major new division in the family tree of planets," according to Andrew Howard, the lead author of the research, "analogous to discovering that mammals and lizards are distinct branches on the tree of life; astronomers like to put things in buckets. In this case, we've found two very distinct buckets for most Kepler planets."[23] The reason for the gap is not clear. It might be that rocky planets can't attract enough material to "jump the gap" and become mini-Neptunes. Or it might be that planets suitable to be in the gap have their atmospheres blown off by their nearby stars, so they "jump the gap" in the other direction.

Let's jump the gap too and turn to the smaller terrestrial exoplanets. Most of these are larger than our planet, and they're called super-Earths. As with mini-Neptunes, super-Earths don't exist in our solar system. I first discuss the crucial role of water in determining a planet's habitability.

10
Water Worlds

"I to the world am like a drop of water that in the ocean seeks another drop, who, falling there to find his fellow forth, unseen, inquisitive, confounds himself."[1] In The Comedy of Errors, *Shakespeare describes the search for kinship. Our ancient ancestors in the tree of life were once all connected by the matrix of the oceans. We emerged tens of millions of years ago and now walk the land as solitary creatures, with the water we carry in our bodies an echo of our origins.*

Water is essential for life, and life as we know it is found everywhere there is significant water. The questions become: What's the minimum amount of water needed for biology to function, and how important is water in determining the possibility of life in exoplanet environments?

Water is the only substance found naturally on Earth as a solid, liquid, and gas. It can dissolve more substances than anything else—even sulfuric acid. It's the third most abundant molecule in the universe. Life on Earth depends on its stickiness and the unusual fact that it expands when it freezes. Every glass of water from your faucet includes molecules that the dinosaurs drank.

"Water is life's matter and matrix, mother and medium. There is no life without water." These are the words of Hungarian biochemist Albert Szent-Györgyi, who won a Nobel Prize for isolating vitamin C and elucidating one of biology's fundamental processes, the citric acid cycle.[2] We take water for granted and yet are acutely aware it's essential for life. Water makes up two-thirds of our body weight; a loss of just 4 percent would result in dehydration, and a loss of 15 percent would be fatal. A person could survive a month or more without food, but only three days without water.[3]

This remarkable compound is a major component of all plants and animals, ranging from 35 percent for some insects to 95 percent for jellyfish. Perhaps one reason we get tired at the end of the day is that we're like water balloons wrapped in flesh, carrying around three five-gallon jugs wherever we go. Water's particular properties let animals regulate their body temperature, and trees pull water up against the force of gravity to their highest leaves. It's a universal solvent, ferrying nutrients and waste products. At a microscopic level, it supports the structure of cells and facilitates the coding of information in the shapes of proteins and the DNA molecule. Water is foundational to the activities of metabolism. In bacterial cells, it's involved in over 99 percent of the biochemical reactions.[4]

How much water is needed for life to survive? Surprisingly, perhaps, the answer is "very little."[5] If we travel to the driest locations on Earth, such as parts of the Atacama Desert where there has been no measurable rain in a century, we'll find microbial life on or near the surface. Similarly for the Antarctic dry valleys, which are the closest to a Mars environment on Earth. Organisms that can tolerate extreme physical conditions are called extremophiles; those that can handle extreme dryness are called xerophiles. Some of these organisms have evolved novel biochemistry to compensate for the lack of water. Others go into suspended animation, with minimal water, and metabolic activity dormant. In one case in the Antarctic, a cell mat that had been dormant for two decades began photosynthesis after a day's exposure to liquid water. Some microbes have even survived the absolute aridity and vacuum of deep space aboard the space shuttle.[6] All branches of the tree of life can manage the trick of going dormant—bacteria, yeast, fungi, plants, and animals.

A few years ago, I visited the Atacama Desert after observing in the south of Chile. A small minibus took me to the vast salt flats an hour south of San Pedro de Atacama. I was surrounded by a surreal white landscape, with the jagged silhouette of the Andes in the east, dominated by the fifty-nine-hundred-meter-high volcano Licancabur. I was standing in a place where it hadn't rained for a century. The white salt crunched underfoot, so I bent down to inspect it. Even in this desiccated environment, I could see hints of lichen. There was life here, clinging tenaciously to the minerals. I found out later that NASA tests its Mars rovers here to give them a serious challenge in searching for life in an arid environment.

Most extremophiles are microbes, but a few are larger. The poster child of extremophile animals is the tardigrade. Boil them, freeze them, crush them, desiccate them, or blast them into space: they survive and come back for more. Discovered in 1773 by German pastor Johann Goeze, tardigrades are known colloquially as "water bears." Up close, they're intimidating, with folds of flesh, eight legs, ferocious claws, and daggerlike teeth.[7] They don't represent a threat, however, since tardigrades rarely grow longer than one millimeter (one-twenty-fifth of an inch) and can only be seen in any detail with a microscope. They've been around for five hundred million years since the first complex forms of life evolved in the oceans. They can lower their water content by a factor of twenty and their metabolic rate by a factor of a hundred. In this desiccated state, they curl up into a little ball called a tun. Tardigrades have been successfully reanimated after a decade of suspended animation. There's even controversial evidence that they can survive for a century. If we were betting on any type of creature that might survive on exoplanets as our first tiny astronaut emissaries, it would be this sturdy extremophile.

Extremophiles expand the envelope of potential habitability. The traditional astronomical definition of habitability is based on a planet's surface temperature being in the range where water can be a liquid. Planets that satisfy this constraint are said to be in the Goldilocks zone of a planetary system. Even life on Earth tells us this definition is too restrictive.[8] Extremophiles are found below the

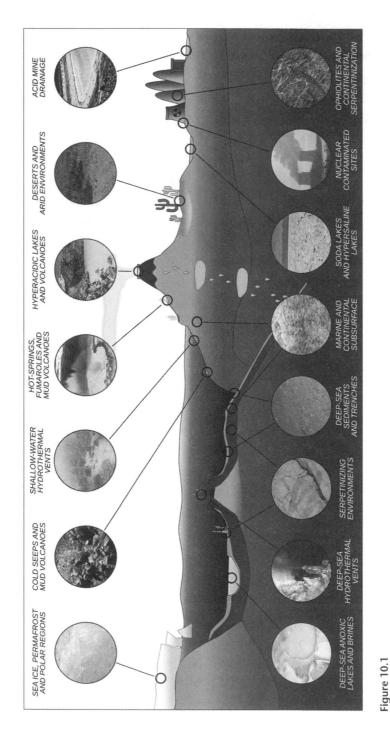

Figure 10.1

Idealized cross-section of Earth's crust showing the diversity of extreme environments where life can be found. This makes it likely that life can exist on many types of exoplanets. (Credit: N. Merino et al., "Living at the Extremes," 2019, *Frontiers in Microbiology,* https://www .frontiersin.org/files/Articles/447668/fmicb-10-00780-HTML-r2/image_m/fmicb-10-00780-g001.jpg.)

freezing point of water, and above its boiling point, near deep-sea hydrothermal vents, far below the surface in mines, and even living in solid rock.

We think of the biosphere as being "painted" on Earth's surface, but startling research in the past few years has shown that life under the ground eclipses life above the ground. By drilling out from the bottom of mine shafts and sinking boreholes into deep ocean sediments, scientists found life everywhere they looked. This subsurface realm has twice the volume of the oceans and holds 10^{30} cells, making it one of the biggest habitats on the planet as well as one of the oldest and most diverse. Some of the life-bearing bodies of water far under the crust have been isolated from the surface for two billion years. Scientists are still puzzling over how life down there survives and thrives. They think that radioactivity splits water into hydrogen and oxygen. Some cells use hydrogen as fuel directly while others get energy from compounds created by chemically reactive oxygen. These remote regions are self-sustaining ecosystems. Welcome to the world within the world.[9]

Our knowledge about exoplanets is still too primitive to know if any of these unusual environments exist around other stars and to what extent. Astronomical methods are poorly suited to detecting life under the surface of an exoplanet. And so with a necessarily broad-brush approach, we consider the likelihood that exoplanets of different types can host surface water and how much. In searching for biology beyond Earth, we obey the mantra of NASA in its exploration of Mars: follow the water.[10]

Sean Raymond has been following the water for most of his life. Growing up in Maine, he experienced many dark and rainy days. Then he went to graduate school in Seattle, where it rains every other day on average. As a graduate student, he made a "splash" with simulations of the formation of our solar system under slightly different scenarios. The computer simulations tracked the rocky and icy particles as planets grew. To this day, no one has worked out why our world is wet. The rocky objects that coalesced to form Earth were too hot to carry much water so the water likely arrived later, from elsewhere. In Raymond's simulations, one to four terrestrial planets formed each time, roughly mirroring our solar system. The water content of those planets varied widely. A few were bone-dry, but most had much more water than on Earth, ranging up to three hundred times the content of all of Earth's oceans. The planets with the most water were twice the mass of Earth: super-Earths. Making a wet planet seemed easy, at least inside a computer.[11]

Since his dissertation, Raymond has made major contributions to ideas about how the solar system and exoplanet systems formed. He also has a well-developed sense of whimsy. He's written about a fifth terrestrial planet that may once have orbited between Mars and Jupiter before a gravitational instability either hurled it into the Sun or out into space. Raymond has speculated about whether moons can have moons. He's written about the various ways that planets can die. These are just the technical articles. On his blog, he indulges in true flights of fancy. There are astronomical poems, culminating in a recent book.[12] There are detailed critiques of science fiction worlds, including a re-creation of Kalgash, the planet dwelling in permanent daytime from Isaac Asimov's story *Nightfall*. His hypothetical planetary systems are wild. "The Solar System is a disappointment," he writes. "It does contain an inhabited planet with forests and oceans and frisbees and beer (Earth). But it only has one. . . . Where are our alien neighbors? Who are we going to space-trade with and have space-wars with or play space-ball games against?"[13] So he rearranges the solar system to fit seven planets and moons into the habitable zone instead of just one. His "Ultimate Solar System" crams six hundred habitable planets into a sixteen-star system. The culmination of his thought experiment is fantastic: a supermassive black hole ringed by a million habitable planets.[14]

Raymond, from the United States, is currently living in Bordeaux, which is a relatively dry part of France—but he retains dual citizenship with Ireland. Perhaps he feels the ancestral tug of an island surrounded by and often doused by water.

As noted in the previous chapter, the real equivalent of Raymond's watery super-Earths makeup about one-third of the more than five thousand known exoplanets, and yet are unknown in the solar system. The other most common kind of exoplanet is the mini-Neptune, two to four times Earth's size. Super-Earths are up to twice the size of our planet and up to eight times its mass. The main difference between these two kinds of exoplanet is that the larger mass and size of mini-Neptunes is due to an outer layer of hydrogen and helium that contributes most of the mass. Liquid water can exist on both mini-Neptunes and super-Earths, but it is the latter, with a rocky surface and shallower atmosphere, where life is most likely to thrive. At what point does a planet stop being mostly rock and transition to being mostly gas?[15]

The key tool for answering that question is sniffing the atmospheres of these planets with spectroscopy (see chapter 2). A planet named GJ 1214b illustrates the importance of spectra for characterizing exoplanets. Discovered in 2009, it is forty-eight light-years from the Sun in the constellation Ophiuchus. This gassy world is 2.6 times the size of Earth and seven times its mass, but only has one-third of its density.[16] The discovery of this first super-Earth was a coup for the young researcher David Charbonneau, who we met previously. He harnessed a fleet of small telescopes no larger than those that amateur astronomers have in their backyards. At the time, GJ 1214b was thought to be a rarity. Now, of course, we know super-Earths are common.

From its overall, average density, GJ 1214b could consist of a rocky core with a hydrogen-helium atmosphere hundreds of miles deep, or a rocky core covered in a deep ocean and atmosphere of steam. Knowing only the overall density, we can't distinguish between these two scenarios. In other words, two distinct planets can have the same mean density. Spectroscopy adds crucial information about the chemical composition.[17] Water can be detected in exoplanet atmospheres, but it's also important to measure carbon dioxide. Together, they can give an estimate of the abundance of heavy elements.[18] These difficult measurements haven't yet resolved the ambiguity in planet structures, but for GJ 1214b the evidence tilts toward it being a water world, made of 25 percent rock, 75 percent water, and a slender sheath of hydrogen and helium. Water worlds overall range from 10 to 90 percent water by mass. (Earth, by comparison, is dry, with just 0.02 percent water in the sum of all the oceans; it's not a water world after all!)

Water is an intimate stranger. We soak in it to relax and clean our bodies, use its solid form to chill our cocktails, and watch its vapor condense into steam from a boiling tea kettle. We take it for granted and forget how essential it is for survival.[19] We're familiar with rivers, glaciers, and steaming hot springs. But to a physicist or chemist, water is more complex and interesting. Take ice, for instance. Almost all ice on Earth—snowflakes, glaciers, icebergs, and the cubes in your drink—are a hexagonal form of the crystal called ice 1h. Scientists, however, have created seventeen different forms of crystalline solid water besides ice 1h—and two forms of noncrystalline (amorphous) solid water—by varying the temperature and pressure beyond the normal ranges found on Earth. The water molecules within the exotic forms of crystalline ice arrange themselves in cubes, tetragons, hexagons, and rhombuses; there's even one form of ice in which the cube collapses to form a two-dimensional square. At extremely low pressures, water molecules can form large empty cages, making the resulting ice half as dense as normal ice.[20] More relevant to exoplanets, at extremely high pressures, the superionic ice discussed in the previous chapter doesn't melt even at temperatures approaching that of the surface of the Sun.[21]

Fantastic fiction can sometimes inadvertently reflect fact. In his 1963 satirical novel *Cat's Cradle*, Kurt Vonnegut describes Ice Nine, a fictional form of ice that freezes at room temperature. If it even touches a drop of regular water, that water will freeze immediately, and the solidification will propagate rapidly in all directions until the whole world is frozen.[22] Vonnegut's Ice Nine doesn't exist (although there *is* a form of ice called ice IX)—but there's something similar. One of the exotic forms of ice that researchers have conjured up in the laboratory, ice VII, can form by rapid nucleation that propagates through the liquid at speeds over 1.6 kilometers per hour (1,000 mph). If the high pressure and temperature conditions necessary to make ice VII exist on an exoplanet water world, that would pose a real threat to its habitability.[23]

Unfathomably deep oceans—water thousands of miles deep, compared to the average ocean depth on Earth of two miles—would exert unimaginably crushing pressures.[24] We'd never survive a descent into the atmosphere of a water world, let alone into its deep oceans—yet let's try to imagine it.

"Water, water everywhere, nor any drop to drink." This line from Samuel Taylor Coleridge's poem *The Rime of the Ancient Mariner* applies to our journey. What we would see depends on how watery the planet is, but also on the distance from its star. Close to the star, the top layer of the atmosphere would be a steamy sauna. Far from the star, the outer layer would be rock-hard ice. In between, we might experience the oddity of an "eyeball" planet, covered in ice and yet with a hole melted on the star-facing side. We descend through the outer vapor and into the murk where light cannot penetrate. Steam turns imperceptibly into liquid, with no well-defined surface. As the pressure and temperature increase, liquid turns into superfluid, then different crystalline ices, and then a plasma phase. Finally, we reach the rocky core. But shockingly, there's no surface. At a pressure approaching a million times Earth's atmospheric pressure, the boundary between rock and water becomes fuzzy. One shades into the other. All the way to the center, the rocks are waterlogged.[25]

The conclusion from these examples is that although water is essential for life as we know it, a watery world is not necessarily habitable and teeming with life. There can be too much of a good thing. Here on Earth, chemical elements vital to life are constantly cycled between the crust, oceans, soil, and atmosphere. Geochemical cycles wouldn't exist on a planet where the rocky core is locked away in a deep icy sarcophagus. Scientists are divided, however, because it might be dangerous to project our Earth-centric perspective onto these alien worlds. Edwin Kite, who studies the habitability of exoplanets, notes, "What I've taken away from this project is the inadequacy of working from Earth's analogy. I love rocks and Earth history, but you really need to build up from basic physics and chemistry, rather than relying on Earth's analogy in order to tackle exoplanet problems."[26]

Avoiding the most massive, water-dominated planets, an argument can be made that other, slightly less watery super-Earths are likely to be superhabitable. Most of them will orbit red dwarf stars, with much longer lives for sustaining life than our Sun. The extra bulk of a super-Earth also helps its habitability. The core stays hot for longer, and tectonic activity replenishes the carbon dioxide, without which the greenhouse effect would fade and oceans would freeze. Here on Earth, an active, molten core sustains a magnetic field that protects us from the perils of the solar wind; there's no reason why the same effect would not occur and be beneficial on a super-Earth. The atmosphere on a larger version of our planet would be thick, and higher surface gravity would increase erosion and make a flatter surface. This sounds boring, but it has an advantage for life. Super-Earths are likely to have many shallow islands, and as on Earth, the biodiversity would be rich in such an "archipelago world." Living there might be tolerable—but dragging around in the stronger gravity would be tiresome and leaving would be difficult. The SpaceX Falcon Heavy rocket can launch twenty-five tons into Earth orbit, but on a super-Earth, whose escape velocity is twice as high, it could barely put a child into orbit.

Recent research has pushed the boundary of habitability from super-Earths up to mini-Neptunes, planets thought to be too massive and hot to support life. Researchers at the University of Cambridge have identified a major category of exoplanets with dense hydrogen atmospheres and large oceans. They dubbed them Hycean planets, from the words "hydrogen" and "ocean." The team leader said, "Essentially, when we've been looking for these various molecular signatures, we've been focusing on planets similar to Earth, which is a reasonable place to start. But we think Hycean planets offer a better chance of finding several trace biosignatures. It's exciting that habitable conditions could exist on planets so different from Earth."[27]

Hycean planets are common in the galaxy, so the nearest and brightest examples will be excellent targets in the search for biosignatures, the spectroscopic detection of any gas that is associated with biology, like oxygen. These ocean-covered planets can support life even if they're significantly farther from their host star than a planet of Earth's size. If they're close to their host star, they'll be tidally locked, and they could support life on their permanent nightsides.[28]

Having made the case for water as essential for life, it's worth pausing to ask if making water a prerequisite for habitability might be too Earth-centric. Perhaps other liquids can act as a medium for biology. A plausible argument has been made by scholar William Bains that other chemistries could be used to build living systems.[29] He has even argued that sulfuric acid could plausibly be a solvent for biology.[30] Astrobiologists from Spain estimated the probability of finding surface seas on rocky worlds composed of nine different liquids, including water. All of these liquids have been detected in the atmospheres of exoplanets and solar system bodies, and all of them are quite abundant in the universe.[31] In order of the decreasing temperatures at which they freeze, they are sulfuric acid, water, hydrogen cyanide, carbon dioxide, ammonia, hydrogen sulfide, ethane, methane, and nitrogen.

The astrobiologists' surprising conclusion: ethane oceans like those on Saturn's moon Titan might be ten times more common than water oceans on planets around a wide variety of stars. Oceans of nitrogen could also be abundant on planets orbiting red dwarfs. We may need to broaden our horizons beyond our favorite liquid.

11

Earth Clones

We descend through the cloud layers to see a rugged coastline. The blue-gray sea below is whipped by a strong wind and dotted with whitecaps. After landing on a rocky plain, we emerge with no protection. The air is bracing and moist, with a slightly acrid taste. We feel unusually light on our feet. Rocks crunch underfoot, and we can see the familiar glint of quartz amid slabs of granite.

The landscape is familiar, but unsettlingly different in the details. The vegetation is tinged blue-green. The leaden sky is brown. We could survive here, but it doesn't feel like home. Is this unfamiliar world as close as we can get to a twin of our planet?

Earth is in peril. Climate change, environmental degradation, tribal conflict, and a population exceeding eight billion threaten our long-term future. Are there any other planets like ours, and other species that share our tribulations? "Methinks you are my glass and not my brother: I see by you I am a sweet-faced youth."[1] Near the climax of *The Comedy of Errors*, Shakespeare describes an encounter between long-separated twins. We imagine that in discovering a true twin of Earth, we'll find life and maybe companionship.

On NASA's dashboard of exoplanet discoveries, planets can be plotted in terms of their masses and orbital periods.[2] About 75 percent have been found with the transit method (chasing shadows) and 25 percent with the radial velocity method (Doppler wobble). Radial velocity planets range from five Earth masses to ten Jupiter masses, with orbital periods from a month to ten years. The sensitivity of that method is too low to find small terrestrial planets.

Transit planets range from below Earth's mass to Jupiter mass, with orbital periods ranging from a few days to a few months. The vast majority of planets are larger than Earth and on more rapid orbits of their stars. The span of data isn't long enough and the quality not good enough to readily identify small planets on one-year orbits.

Among the more than five thousand exoplanets so far discovered, we've not yet found a clone of Earth—by which we mean a planet with Earth's mass and size on a yearlong orbit around a yellow, middle-aged, main-sequence star.

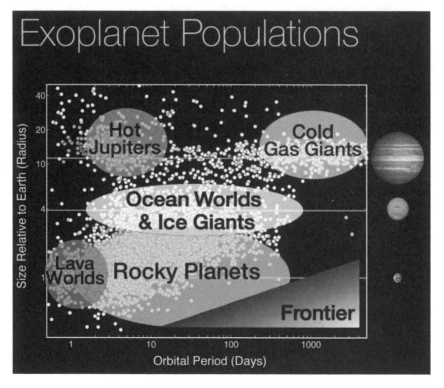

Figure 11.1
The population of exoplanets, as discovered mostly by the radial velocity (top half of the diagram) and transit (bottom half) methods. True Earth clones lie in the bottom-right corner of the diagram. (Credit: NASA/Ames Research Center/N. Batalha/W. Stenzel, "Exoplanet Populations," last updated 2017, https://www.nasa.gov/image-feature/ames/kepler/exoplanet-populations.)

If we relax the requirement that the planet orbit a star like the Sun, however, there are several near twins of Earth. Most stars are less massive than the Sun. Red dwarfs are from 10 to 50 percent of the mass of the Sun (below 8 percent, a star does not heat up enough to release energy by fusion and shine). Red dwarfs live for hundreds of billions of years, and there are a hundred red dwarfs for every star like the Sun.[3] Twenty of the thirty stars nearest to the Sun are red dwarfs, including the closest, Proxima Centauri. Red dwarfs have slender habitable zones that are much closer than the habitable zones of sunlike stars are (because the star is cooler than our Sun). Nevertheless, the vast number of red dwarfs means that their habitable "real estate" is greater than the habitable real estate around stars like the Sun. Other factors weigh against them as places where life might develop. Red dwarfs emit X-rays that would irradiate any planet nearby. Planets orbiting red dwarf stars are tidally locked to always have the same face pointing to the star, creating large temperature gradients from the daytime side to nighttime side. The habitability of red dwarf planets is hotly debated, but researchers think their surface conditions may allow life.[4] Their numbers plus their long lifetimes mean that finding life on red dwarf planets is hundreds of times more likely than finding life on the planets of solar-type stars.

Anticipation started building after discoveries made by the High Accuracy Radial Velocity Planet Searcher (HARPS), a Doppler spectrograph attached to a 3.6-meter telescope at ESA's La Silla Observatory in Chile. In 2012, the HARPS team, led by Michel Mayor, announced the results of their survey of a hundred red dwarfs over six years. They found nine super-Earths, one to ten times Earth's mass, with two in the habitable zones of their stars. "Our new observations with HARPS mean that about 40% of all red dwarf stars have a super-Earth orbiting in the habitable zone where liquid water can exist on the surface of the planet," said team member Xavier Bonfils. "Because red dwarfs are so common—there are about 160 billion of them in the Milky Way—this leads to the astonishing result that there are tens of billions of these planets in our galaxy alone."[5] If there are super-Earths around red dwarfs, surely there are Earths as well.

In 2015, a spectacular system was discovered around TRAPPIST-1, an ultra-cool red dwarf forty light-years away in the constellation of Aquarius. The transit technique allowed astronomers to push down to the regime of earthlike planets. Three terrestrial planets were found initially, with four more announced in 2017.[6] All seven planets are truly earthlike, ranging from 0.4 to 1.4 times Earth's mass, and three are in the habitable zone of the red dwarf. Since they orbit a star nothing like the Sun, we can call these planets Earth cousins rather than Earth clones. Their orbital periods range from 1.5 to 19 days, and all are much closer to their star than Mercury is to the Sun. There's evidence for water in the atmosphere of one of them.

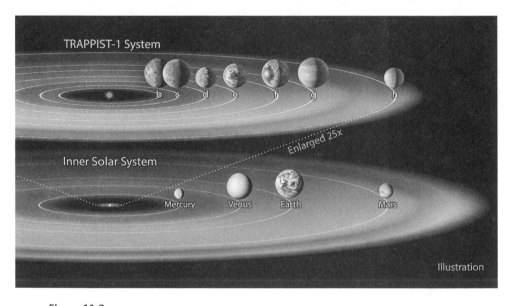

Figure 11.2
The red dwarf star TRAPPIST-1 has seven earthlike planets, all closer to their star than Mercury is to the Sun. Planets e, f, and g are in the habitable zone of the cool star. (Credit: NASA/JPL-Caltech, "TRAPPIST-1 Habitable Zone Compared to Our Solar System," 2018, https://www.planetary.org/space-images/habitable-zone-in-trappist-1.)

The TRAPPIST system is another stunning example of Kepler's harmony of the spheres: the inner six planets form the longest-known resonant chain, with orbital ratios of 8/5, 5/3, 3/2, 3/2, and 4/3. Recent work suggests that all the planets have similar compositions to each other, but they are all different from Earth.[7] They're less dense, and have less iron and substantially more water. TRAPPIST-1 is a compelling target for future observations to sniff the planetary atmospheres and look for their alteration by biology.

Let's imagine the view from the most earthlike of these planets. The small star looks like a blood orange in the sky and bathes the scene in ocher light. Three of the other planets are suspended in the sky, each appearing larger than our Moon. A year is a week in this planetary system, so the planets move perceptibly from moment to moment. It would be an arresting but alien scene.

The Kepler mission identified a dozen earthlike worlds orbiting in the habitable zones of red dwarf stars. Now the new kid on the block is getting in on the act. In 2021, TESS (see chapter 6) found its first Earth cousin: a rocky planet almost exactly the size of Earth orbiting a red dwarf a hundred light-years away.[8] This planet gets from its star 90 percent of the energy Earth gets from the Sun. It was just one of the goodies in a release of twenty-two hundred candidates from TESS in early 2021.[9] Astronomers are like kids in a candy store as they sift through the data and pick objects for further study. Dozens more habitable earthlike planets are anticipated.

So far, the closest to an Earth clone is Kepler 452b. Technically, it's a super-Earth at one and a half times the size and five times the mass of Earth. It orbits its sunlike star in 385 days and at a distance of 1.04 AU, almost perfectly matching our orbit. It receives energy from its host star at a 10 percent higher rate than we do from the Sun and is considered to be habitable.[10] We've no idea if there's life on Kepler 452b, but its star is two billion years older than the Sun, so there has been plenty of time for evolution to progress from chemistry to biology. Traveling there isn't an option. At a distance of 1,400 light-years, it would take our fastest space probe, Voyager 1, over 10 million years to get there (and Voyager 1 is not leaving our solar system in the right direction). But radio astronomers searching for extraterrestrial intelligence are scanning nine billion frequency channels for radio transmissions from this star system (for more on the search for extraterrestrial intelligence, see chapter 20). To date, ET has not spoken.[11]

Most exciting of all, the star nearest to the Sun has a possibly habitable planet. Proxima Centauri is a red dwarf 4.4 light-years from Earth, three hundred times less than the distance to Kepler 452. There had been hints of a planet since 2000, but in 2016, a European team started an intensive observing campaign that it called "Pale Red Dot." The planet that the team found is 10 percent larger than Earth and at least 10 percent more massive. It orbits in 11 days, and the average surface temperature is 20°C (36°F), slightly chillier than Earth's.[12] There's also a super-Earth orbiting much farther out, beyond the habitable zone, and a hint of a Mars-mass planet closer in that might be habitable.[13] It gets better. Proxima Centauri is part of a triple system, well separated from Alpha Centauri A and B, which orbit each other every 80 years. In 2021, astronomers used infrared imaging to detect a planet with a mass between Neptune and Saturn orbiting Alpha Centauri A. It has an orbit quite like Earth's, but is unlikely to be habitable.[14]

An innovative new telescope is planned to spot planets around Alpha Centauri A and B. Toliman is the ancient Arabic name for Alpha Centauri, but it also stands for Telescope for Orbital Locus Interferometric Monitoring of our Astronomical Neighborhood. The goal is for this shoebox-size telescope to go into Earth orbit in 2023. It will find Earths by the most challenging method of all: astrometry (see chapter 7). But rather than trying to make a high-resolution image, the telescope will spread the light from the stars over thousands of pixels, making a photon "fingerprint" of each star's spatial position on the sky. The goal is to see minuscule shifts in either star's position caused by the gravitational tug of orbiting planets. The astrometry is made easier because there are two stars instead of one. Project leader Peter Tuthill, of the University of Sydney, says the approach is "a bit of an optical trick. . . . Alpha Centauri is just a sitting duck for this particular technique. It's almost like the universe put it there for this particular mission."[15]

This system offers a bonanza since habitable planets can be detected around three stars at once. Its proximity means the sensitivity to imaging Earths is better than anywhere else. It's even close enough that we might one day visit, as I'll discuss later.

Now let's get quantitative. In 2011, astronomers proposed an "Earth Similarity Index," which is designed to characterize how similar a planet or moon is to Earth. It has a scale that runs from 0 to 1, where a perfect clone of Earth has a value of 1. The index incorporates the planet's size, density, surface temperature, and escape velocity.[16] It doesn't calculate planet habitability or the ability of a planet to host life—just the similarity of its bulk properties to our planet. Using this scale, Mars has an index of 0.64. Mars is smaller than Earth yet may have subsurface life, so it's considered to be a planet on the "edge" of habitability. A database of potentially habitable planets lists forty-eight with an index higher than Mars. Nine of these planets have indexes of 0.85 or higher, including two in the TRAPPIST system, and Proxima Centauri b.[17]

Although we haven't found an Earth clone yet, it almost certainly exists. The exoplanets detected so far are almost all within four thousand light-years. That's just 1 percent of the Milky Way, which is one hundred thousand light-years across. With tens of billions of terrestrial planets in the galaxy, probability and common sense say that many will be like Earth.[18] Multiply by a hundred billion galaxies in the observable universe, and the projection becomes ten billion billion (10^{19}) terrestrial planets around sunlike stars and fifty times more around red dwarfs.[19] So the nearest exact twin may be on the far side of the galaxy or in another galaxy millions of light-years away. Finally, if we embrace the speculative multiverse theory, where our entire observable universe is just one region in a potentially infinite ensemble of universes, not only are there many exact copies of Earth, but there are many exact copies of you and me![20]

Leibniz claimed that we live in the best of all possible worlds. We can look for Earth cousins and Earth clones, but is our planet really the gold standard when it comes to defining habitability? Maybe you like your parents, if they fed you, housed you, and rarely made your life miserable. You might even think they are great parents. But to properly judge them, you would have to find out more about other parents. It's statistically unlikely that they're the best of all possible parents. The same goes for planets and habitability.

Rather than being the best of all possible worlds for life, Earth might be inferior. Life is found almost everywhere there is water, but it has trouble thriving in arid deserts, frigid polar regions, and nutrient-poor oceans. Biodiversity is highest in the equatorial regions, especially rain forests.[21] The traditional definition of a habitable zone is flawed. Even when a rocky planet is in the habitable zone, it might lack the atmosphere needed to shelter life or the geological activity needed to drive it. Conversely, tidal heating can render moons habitable in the frigid outer parts of a solar system. Jupiter's Europa is the perfect example. Researchers have come up with a list of the attributes that make a planet superhabitable.[22] They include a mass twice that of Earth and a size 20 to 30 percent larger, oceans that are shallow enough for light to stimulate life below the surface, an average temperature of 25°C (77°F), an atmosphere thicker than Earth's but with more oxygen, and a parent star older than the Sun yet younger than seven billion years. A list of two dozen exoplanets more habitable than Earth using these criteria was published in 2020.[23]

Habitability is a moving target. As we saw from the vignettes in the prologue, Earth has suffered extreme conditions and catastrophic changes in the distant past. Things could have turned out very differently for our planet if the initial conditions had been even slightly different. Just as cloned plants, with identical genetic material, can become distinct if they grow up in different environments, two identical planets can develop quite differently, with small differences magnified over time. Imagine two identical Earths where a slight difference in the star's radiation alters the evolutionary clock by 10 percent. After 4.5 billion years on one, you'd find plants, animals, and us. While after 4.5 billion years on the slightly cooler Earth, you'd seen barren land and no life in the oceans larger than the head of a pin.

Lisa Kaltenegger has thought hard about habitability. One of a younger generation of women leaders in astrobiology, we met her earlier with her list of planets that have the right vantage points to detect life on Earth. Born in Austria, she's always had a wide range of interests: "If I weren't a scientist or an engineer, I probably would make movies or paint or be a photographer." Kaltenegger approaches her subject with humor and a light touch. She knows that water worlds are abundant, and the cold ones will be easiest to detect: "Frozen-over planets should pop out in our search like that because they're so much brighter and reflect so much more light." But water worlds with no continents have an appeal as well: "You'd have an ultimate surfing destination. If you ever wanted to go there you can surf, surf, surf, as long as you don't want to go ashore."[24]

Kaltenegger and her colleagues at Cornell have run models of Earth's atmosphere over its history, and then used them to generate an atlas of spectra at different times. These models form a guide for researchers planning to sniff atmospheres of exoplanets for signs of life. Kaltenegger describes the motivation: "Using our own Earth as the key, we modeled five distinct Earth epochs to provide a template for how we can characterize a potential exo-Earth—from a young, pre-biotic Earth to our modern world. The models also allow us to explore at what point in Earth's evolution a distant observer could identify life on the universe's pale blue dots and other worlds like them."[25]

Around 4 billion years ago, Earth's atmosphere held much more carbon dioxide, methane, and ammonia than it does now. Earth was a water world, with oceans high enough to submerge today's continents above the level of Mount Everest.[26] There was no oxygen, so this early Earth would have been fatal to us. Life had probably already started, deep in the oceans—although it would be almost impossible to detect from afar. By 3.5 billion years ago, carbon dioxide had declined and the atmosphere was predominantly composed of nitrogen. Three subsequent epochs trace the rise of oxygen from 0.2 percent to its current value of 21 percent. "Our Earth and the air we breathe have changed drastically since Earth formed 4.5 billion years ago," explains Kaltenegger. "This paper addresses how astronomers trying to find worlds like ours could spot young to modern Earth-like planets in transit using our own Earth's history as a template."[27] Remote observations of Earth would only show a biosphere starting 2 billion years ago.

Far from being inevitable, the habitability of Earth has always been "touch and go." The climate has veered from ocean-boiling heat to planet-wide deep freeze. Simulations suggest the long-term habitability of our planet wasn't inevitable but instead contingent.[28] It could have gone either way, and we're literally lucky to be alive. Living Earths might be much rarer than we hope.

12
Exomoons

"It is the very error of the moon; she comes more nearer earth than she was want, and makes men mad."[1] *As alluded to by Shakespeare, it might be lunacy to consider life on a moon.*

We've seen that the detection of exomoons presents an extreme challenge. Assessing their potential habitability will prove even more difficult. And yet exomoons might well be hosts to basic microbial life, or even advanced organisms, since not all moons are barren and airless like our own natural satellite. If other planetary systems are like ours, then there are more exomoons than exoplanets and there could be more life on moons than there is on planets.

Since the data on exomoons is so sparse, it makes sense to view them through the lens of the moons in our solar system. Altogether, there are just over two hundred moons in orbit around the eight planets.[2] Moons are sparse in the inner solar system. Mercury and Venus are moonless, Mars has two small, potato-shaped moons, and we have our familiar Moon that from our position, appears to wax and wane monthly in the sky. Mars's gravity probably captured its moons, Phobos and Deimos, from the asteroid belt, so they're as uninhabitable and lifeless as our own airless Moon.[3] But we should get past any thought that all moons in our solar system are boring rocks in space.

Shifting our attention to the outer solar system, the population rises steeply. Jupiter and Saturn have about eighty moons each, and Uranus and Neptune forty moons between them. If our solar system is typical, the vast majority of exomoons will be orbiting gas giants and ice giants—and they will be anything but boring.

A few examples illustrate the diverse attributes and personalities of the moons of the outer solar system, allowing us to envisage what possibilities might be out there among the moons around exoplanets.[4] There are sulfurous, pockmarked moons. There are moons with volcanoes, some sending lava dozens of miles high, and others spewing slushy ice and even liquid nitrogen. Some moons are tiny, but many are surprisingly large; one is even more massive than Mercury and Pluto. There are moons with deep subsurface oceans. One moon has a magnetic field, and another has lakes of liquid ethane and methane under a veil of hydrocarbon smog.[5]

These moons are our neighbors, but we know surprisingly little about them. We've sent fifty missions to Mars, but only nine to the outer solar system. All nine visited Jupiter, and four visited Saturn. Yet only one probe, Voyager 2, visited Uranus and Neptune—and that was over thirty-five years ago.

What is it that makes a moon interesting and potentially habitable? Normally, habitable zones are defined simply by the distance from the star, with energy received at a high enough rate for water to exist as a liquid on the surface of the planet or moon. In the outer solar system, the Sun's energy is feeble, twenty-five times less than Earth receives at the distance of Jupiter and sixteen hundred times less at the distance of Neptune. That far from the Sun, there can be no oceans at the surface of a moon. Liquid water can't exist on the surface of an extremely cold object with little to no atmosphere. Either it will be frozen or it will simply boil away. Mars is a good example. Much of the surface temperature is far below freezing; the water there only exists as solid ice. In some places, the temperature is as high as 20°C (70°F)—but because of the low atmospheric pressure, the ice simply turns straight into a gas (it sublimes). If instead, you put a glass of water on these warmer parts of Mars, the water will boil away in a few seconds. That seems crazy, but at an extremely low pressure, the boiling point of water is lowered. Just refer to a high-altitude cookbook if you need convincing. On mountaintops, where the atmospheric pressure is low, water boils at a lower temperature than at sea level.[6] At twelve thousand feet, a three-minute boiled egg takes six minutes to fully cook. Mars just takes high-altitude cooking to an extreme—and it would be more dramatic on a low-mass moon, whose atmosphere would be even more rarefied.

The atmospheric pressure of a moon may be extremely low, but there could still be liquid water under the moon's surface, such as perhaps a lake or ocean under the crust. This is true even for moons that are far from their host planet's host star. For this to be possible, a source of heat is needed. Any moon can get energy in two ways.

One is from its mass of rocks. Small moons are geologically dead while large moons have enough internal energy to be interesting and keep water liquid. Radioactive elements like uranium and thorium occur naturally in rock, and their long decay times mean they keep generating heat for billions of years. When the moon is small, that heat leaks into space, and the moon will cool off; think of a small stone taken from a fire pit, it will cool off much quicker than a large stone from the same firepit. A moon larger than a thousand kilometers (six hundred miles) across has a heat diffusion time longer than the age of our solar system, so it has a reservoir of energy to maintain a subsurface ocean.[7]

The second source of energy is tidal heating. Tidal forces physically distort moons that are close to a massive planet because the gravitational attraction to the host planet is greater on the near side of the moon than on the far side. If the orbit is elliptical, the tidal force changes through the orbit, so the moon is getting alternately stretched and squeezed. Think of squeezing a squash ball or racquetball over and over. It will gradually heat up. In some moons, this form of heating can be greater than the heating from radioactive decay.

These two sources of heat provide the energy for the exotic effects described in the moons of the outer solar system above—and they maintain a temperature high enough for water to remain liquid. The right combination of mass, tidal heating, and protection by a rock and ice outer layer can sustain water liquid far from the Sun. The shell of rock is an insulating blanket, and pressure from the rock helps keep it liquid. Salt lowers the melting point too. Briny or salty water is liquid down to −25°C (−13°F), and some combinations of salts and water stay liquid down to −60°C (−76°F). Add ammonia and the mixture stays liquid down to a frosty −100°C (−148°F).

Because of these combined effects, massive moons are more likely to have active geology, liquid water, and possibly life. No exomoons have been detected, but our understanding of how moons form allows us to predict their properties. The four giant planets in our solar system each have moon systems that add up to 0.01 percent of the planet's mass. This regularity is a result of two competing processes during formation: the supply of inflowing gas to form the moons, and loss of moons due to orbital decay (spiraling in toward the planet) caused by the gas.[8] The same scaling might apply in other planetary systems. So since many of the gas giant exoplanets are larger than Jupiter, we can reasonably expect that many exomoons could easily be as massive as our Moon and some might be more massive than Mars. These beefy exomoons are the most likely to be habitable.[9]

The best candidates for hosting earthlike moons will be large Saturn- or Jupiter-mass exoplanets orbiting within the habitable zones of their stars. This is another Goldilocks zone since hot Jupiters are too toasty and gas giants like those in the solar system are too frigid. Kepler data has been used to identify over one hundred suitable exoplanets, and they will be the subject of intense scrutiny in the coming years.[10] Astronomers think there may be more habitable exomoons in the galaxy than habitable exoplanets. This projection is based on our solar system, which has one habitable planet—Earth—and a dozen moons of the giant planets with liquid water and energy that could power familiar or strange forms of biology. We like to think of our home as the "water world," but several of those moons have much more water than Earth.

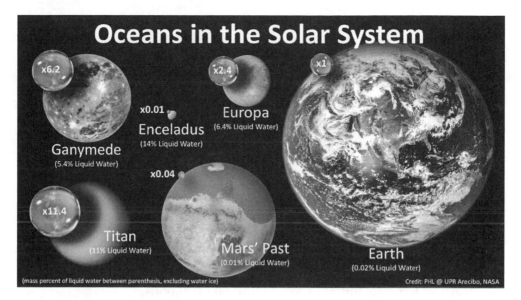

Figure 12.1

Oceans in our solar system. Several moons of the gas giant planets have more liquid water than all of Earth's oceans—something that is also likely to be true of exomoons. (Credit: Planetary Habitability Lab, University of Puerto Rico, in "Oceans of the Solar System" by M. Özgür Nevres, 2020, *Our Planet*, https://ourplnt.com/wp -content/uploads/2020/10/Oceans-Solar-System.jpg.)

There's another kind of exomoon that deserves our contemplation. Long before *Avatar*, filmgoers encountered a moon with endless forests called Endor as the setting for the pivotal battle in *Return of the Jedi*. Sean Raymond, with his sideline in adjudicating the science in science fiction, was drawn into a discussion on the fate of the cute but cannibalistic Ewoks as the Death Star exploded. "If a chunk of the Death Star escaped the gravity of Endor," Raymond decided, "a corresponding chunk was kicked in the other direction and must have killed the Ewoks."[11] He pointed out that as a satellite of the Death Star, Endor was a submoon. This thought led him to write a scientific paper about the possibility with a colleague.[12] They concluded that submoons could exist, and in our solar system they might be found around Titan, Callisto, or our own Moon.

Raymond took it one step further: "It's hard to imagine a planet hosting subsubmoons—moons of submoons—because everything gets closer and closer together and tides are even stronger. That means only teeny, tiny subsubmoons could exist."[13] If by now you're getting dizzy, let me reel this in by noting that there's still one viable candidate for an exomoon: a Neptune-size object orbiting the super-Jupiter exoplanet Kepler 1625-b.[14] This exomoon is big enough that it could potentially host an earthlike submoon![15]

Let me finish our consideration of the habitability of exomoons by returning closer to home and visiting a few of the dozen or more habitable locations that are far beyond the traditional Goldilocks zone—in locations where water cannot exist as a liquid on the surface of a planet. These large moons of the gas giant planets give us a taste of what we might expect to find on exomoons.

We're orbiting the fourth largest of Jupiter's moons. Below us is the smoothest and shiniest object in the solar system. As we approach the brilliant white surface, a filigree of delicate dark streaks becomes visible. The streaks crisscross the satellite, which is named after the consort of Zeus, who abducted her disguised as a white bull. This is Europa, and it's big—almost the size of the Moon. It has twice the water of all of Earth's oceans.[16] Closing in, more features come into view: ridges of dark material, domes and pits, and irregular shapes, like icebergs trapped in a frozen sea. The surface looks young, as if it has melted and frozen many times, water upwelling through crevasses in the ice, only to freeze instantly on meeting the vacuum of space. Standing on the surface, it's −160°C (−256°F), and the ice is hard as granite. The smooth surface is punctuated by a forest of slender ice spires, extending fifteen meters high. The discoloration on the surface is caused by clays, magnesium, and sulfur salts. In the distance, a vent spews ice crystals a hundred miles into the sky. There's no air to carry sound, but we imagine the ice heaving and cracking under our feet.

On Europa, we're in the tight gravitational grip of the huge gas giant planet Jupiter, which looms in the sky like a vengeful god, its

Great Red Spot a bloodshot eye. The radiation is intense—strong enough to provide a lethal dose in a day. We enter our bathysphere and wait as the nuclear plant slowly melts through the ice. We've chosen a place where the "chaos" of ice is less than a mile thick; in other locations, it's twenty or thirty miles thick. After hours of descent, we're released into the gloom of a salty ocean. The water temperature is kept tepid by radioactive decay from the moon's rocky core and the relentless tidal squeezing of Jupiter. The ocean floor is unfathomably far away, over two hundred miles below us. Could there be life here?[17] It's difficult to tell, but bacteria and algae might be clinging to the ice above our heads, and hydrothermal vents are scattered over the seafloor, like black smokers in Earth's oceans. It was dynamic interactions between rock and seafloor that gave rise to life on Earth four billion years ago.

Not far from our previous stop, we approach the largest moon in the solar system, almost half the size of Earth. It's named after the most beautiful of mortals, Ganymede, who was abducted by the gods to serve wine to Zeus. But Ganymede the moon is austere and forbidding, with a dark surface pockmarked by craters and expanses of dirty ice. It's locked in an orbital resonance with Europa, completing one orbit of Jupiter for every two by Europa. Under the crust, there's a salty ocean that's the largest in the solar system.[18] It has an unusual structure, like a "club sandwich," with alternating layers of water and ice. The outer layer is ice 1h, like the ice in your cocktail, but lower down, under increasing pressure, are layers of different kinds of ice, including ice VI. Intense compression forces the ice crystals into more compact configurations.[19] The Galileo spacecraft swooped to within 240 kilometers (150 miles) in 1995, but we can only speculate about the subsurface scenery.

Moving on, we venture even farther from the Sun to the largest moon of Neptune. At the surface, we're standing on a frozen lake of translucent ice with darker material visible through the surface layer. But this is not water ice. It's frozen nitrogen. The temperature is 250°C (−420°F), cold enough to freeze air. We're on Triton, so far away from the Sun that our star appears a thousand times fainter than it does from Earth. Triton's tilted and backward orbit suggests it didn't form here but instead was captured from the realm beyond the planets. Neptune, god of the sea, with Triton, his captive son carrying a conch shell he blows like a trumpet. Geysers shoot nitrogen jets miles into the sky, then falling and dusting the surface with a fresh layer of nitrogen snow. Despite the numbing cold, heat trapped under the surface keeps water liquid in a vast subterranean ocean.[20] We're in a habitable zone 4.5 billion kilometers (2.8 billion miles) from the Sun.[21]

Even farther out, we find ourselves on the surface of Pluto. Pluto is not a moon, but it's not a planet either. It's one of several large objects in a region of the solar system that's crowded with millions of small objects, called the Kuiper belt. Pluto has a looping orbit that takes it as far as 7.4 billion kilometers (4.6 billion miles) from the Sun. The god of the underworld has been banished from the pantheon of planets and demoted to the status of a dwarf planet. Sunlight that takes just eight minutes to reach Earth takes over five hours to reach Pluto, illuminating us feebly. As was the case on Triton, we're once again standing on nitrogen ice. Within it, convection cells circulate floating blocks of water ice. There's a water ocean a hundred miles deep far below our feet. When the New Horizons spacecraft sped past Pluto in 2015, it transformed our understanding of this small world.[22] This dwarf planet has active geology, a young surface, copious amounts of water, and all the ingredients for life.

We've reached the bone-chilling outer reaches of the solar system, but there's still habitability here and a chance of biology. For our last stop, we plunge back in toward the Sun to find the moon that's perhaps the most likely to host life. This is a place the Cassini mission flew by more than forty times, and where the Huygens probe landed in 1995 and beamed back ninety precious minutes of data before its battery died.

We descend through an atmosphere that is an orange-tinged haze. It's like landing in Los Angeles on a smoggy summer day. As we near the surface, we test the air and find it's mostly nitrogen, as on Earth, but thicker. Add some oxygen and it would be breathable. The surface comes into view, and we see familiar topography: mountains, valleys, river deltas, lakes, volcanoes, and glaciers. As we touch down, a gentle rain is falling. Testing our weight, it's one-seventh of what it would be on Earth, similar to the Moon's gravity.

Exiting the spacecraft, any sense of familiarity ebbs away. It is bitterly cold, −180°C (−290°F). The nitrogen in the air is harmless, but the other gases are toxic: methane, acetylene (ethene), propane, hydrogen cyanide, and carbon monoxide. Just as well there isn't any oxygen. If there was, all you'd have to do is light a match to get a conflagration. We're in what looks like a riverbed, but the smooth pebbles aren't rocks, they're made of frozen methane and ethane. The lakes are filled with those liquids too, and their surfaces are dark, almost black. There are dunes, but instead of sand, their rippled surfaces are made of organic solids and frozen hydrocarbons. A fierce wind whips the light material into a dust storm that ascends high into the ocher sky.

The volcanoes are alien too. Rather than molten rock, they're ejecting ice particles and ammonia. This is not a welcoming place. Before we dropped down into the atmosphere, Saturn loomed in the sky, a hand's width at arm's length. We're nearly a billion miles from home. Welcome to Titan.

If there's life here, it's unlike anything on Earth. Water is the medium for biology on Earth. On Titan, the medium would be liquid hydrocarbons. Titan offers a test case for the ubiquity of life in the universe. Planetary scientist Jonathan Lunine describes it this way: "Should the methane-ethane seas of Titan host an exotic form of life, one not based on the same biochemistry as that in water on Earth, the question of whether or not life is a common cosmic phenomenon will largely have been answered."[23]

13

Rogue Planets

"It goes so heavily with my disposition that this goodly frame, the earth, seems to me a sterile promontory. This most excellent canopy the air, look you, this brave o'erhanging, this majestic roof fretted with golden fire—why, it appears no other thing to me than a foul and pestilent congregation of vapors."[1]

In his melancholy mood in the quote above, Hamlet offers a bleak view of Earth and humanity. Declaring humans to be "the quintessence of dust," he anticipates the twentieth-century realization that we're made of stardust. Earth tossed from its star system, floating the dark seas of interstellar space, would seem a bleak place to hope for life. More likely it would be a ghost ship, intact and under sail, with its crew dead or missing, like the Marie Celeste.[2] *Or would it?*

I conclude my inventory of habitable worlds by looking at orphans with no parent star. Sometimes called "nomads," suggesting a restless spirit, or "rogue planets," suggesting a naughty disposition, they might be the submerged part of the metaphoric exoplanet "iceberg," and so actually be the dominant population.

To get to the radical proposition that planets can be uncaged, roaming through space without a star, we must investigate the stability of planetary orbits. To the casual eye, our solar system is stable. In Shakespeare's soothing words, the planets "observe degree, priority, and place."[3] Tidal interaction with the Sun increases Earth's distance from it by a millionth of a meter (0.00004 inches) per year. As the Sun consumes nuclear fuel, its mass decreases (since it radiates energy, and by Einstein's $E=mc^2$, it loses a small amount of mass)

CREDIT: PHL @ UPR Arecibo (phl.upr.edu)

Figure 13.1
Types of habitable worlds. Planets can be in the traditional habitable zone, moons can have liquid water underground far from the Sun, and nomads ejected from their solar systems could have thick atmospheres and surface water. (Credit: Planetary Habitability Lab, University of Peurto Rico, "Three Types of Habitable Worlds," 2012, https://phl.upr.edu/library/media/three-types-of-habitable-worlds.)

and its grip on Earth steadily loosens. That increases our distance from the Sun by twenty millimeters (seven-eighths of an inch) a year.[4] Even over millions of years, these are miniscule effects, and they'll never be sufficient to cause the ejection of our planet from the grip of the Sun's gravity.

But in the 1980s, when computers became powerful enough to simulate the formation of planets around stars, astronomers noticed something interesting. Early in the history of a solar system and much later in its evolution, the behavior of planets is wildly unpredictable. Motions are chaotic. This is chaos in the technical sense of mathematical chaos theory, where a small change in the initial conditions can vastly change the outcome and long-term predictions are uncertain.[5] In 1988, a group working at MIT used a supercomputer to integrate the orbits in our solar system for over a billion years. They found that Pluto's orbit shows signs of chaos.[6] That means the whole solar system is potentially chaotic since each planet affects all the others, however slightly, and those gravitational nudges pile up over time. Another group simulated twenty-five hundred possible futures for the solar system. Varying Mercury's position by as little as a meter (three feet) could make it collide with Venus or plunge into the Sun. An error as small as fifteen meters (fifty feet) in measuring the position of Earth today makes it impossible to predict where Earth will be in its orbit a hundred million years from now.[7] This is the "butterfly effect" made popular by mathematician Edward Lorenz. Fifty years ago, he speculated that the time and location of a tornado might be affected by a butterfly flapping its wings far away and weeks earlier.[8]

I learned how regularity can slip into chaos in the physics lab when I was an undergraduate student in London. The experiment was a dripping faucet. At first, it seemed too mundane to be worthy of study; I'd been hoping for something involving high-energy particles or strong magnetic fields. With the tap wide open, the water came out in a steady stream. When it was nearly cinched off, water came out as a metronomic and annoying series of drips. But in between, something interesting happened. Just before the drops merged into a single stream, their rate seemed to ebb and flow. It was like they fell with a stutter step. Using a strobe light confirmed this behavior. Their motion frozen by the strobe, I could see the frequency of drops varying erratically. Something as simple as a dripping faucet was a complex system on the edge of chaos.[9]

The early solar system was chaotic in the normal sense of the word too. Planets and asteroids carry scars from collisions, and some planets migrated inward and outward soon after formation. Hot Jupiters are evidence for the wholesale rearrangement of the "furniture" in distant solar systems. Sean Raymond has been modeling planet formation for his whole career. He believes our solar system is weird—a one-in-two-thousand rarity. The simulations show that instabilities among giant planets are extreme, with the gravitational slingshot effect ejecting one or more planets into interstellar space. In some of the simulations, the ejected planets outnumber captive planets. Raymond says, "Planetary formation is a dynamic process, marked by big instabilities and lurching movements. No wonder there is such a dizzying array of systems out there."[10]

In 2011, David Nesvorný of the Southwest Institute in Boulder, Colorado, ran thousands of simulations of the orbits of planets in the young solar system.[11] His work suggested that there were probably originally five giant planets, not the four we have today (Jupiter, Saturn, Uranus, and Neptune), and one of them was ejected early on. The stability of the present-day solar system, including our privileged position in the Goldilocks zone, might be due to the jettisoning of a huge planet at an early point—a planet that's now wandering in the vast depths of space at high speed. Even into its dotage, the solar system won't be sedentary. As stars drift randomly through the galaxy, their paths sometimes cross. Unlike star-crossed lovers, they don't have to touch or kiss because gravity's long reach means they cause mayhem just by proximity. Stars sideswiping the solar system will typically eject one giant planet after thirty billion years and all of them after a hundred billion years.[12]

Planets without a star are nomads or rogue planets. Since exoplanets are primarily found by their effect on the parent star, either tugging or dimming it, how do we discover exoplanets with no nearby star?

The trick is to use microlensing, explored in chapter 5. The rogue planet happens to pass in front of a more distant star, and the starlight bends and is amplified as described by general relativity. Einstein worked out the theory of gravitational lensing in 1936, but he set it aside, concluding that "there is no great chance of observing this phenomenon."[13] Luckily, he was wrong, although it wasn't until 1979 that lensing was first detected. As we've seen, microlensing has been used to find around a hundred exoplanets, yet almost all of them are bound to stars. The brightening of the background starlight due to a passing planet is brief—days for a Jupiter-mass planet and hours for an Earth-mass planet.[14]

There are two dozen rogue candidates from microlensing surveys, and only a handful have been confirmed. The most exciting of the few detections was found by sifting through twenty years of faint stellar flickers from a Polish microlensing experiment.[15] It's an object with a mass between that of Mars and Earth. In two decades of data, the signal rose and fell for just forty minutes. Blink and you'd miss it. This one object gives a taste of the population of free-floating Earths that might be out there. Finding many more will depend on the 2.4-meter Nancy Roman Space Telescope (see chapter 6). Using microlensing, that telescope is expected to discover several thousand exoplanets, a few hundred of which should be rogue planets as small as Mars.[16]

At the end of 2021, a team of astronomers from the European Southern Observatory announced the largest batch of rogue planets yet, and at least seventy with masses comparable to Jupiter's. They used imaging techniques to see the planets directly in a nearby dense star-forming region, as opposed to the indirect and fleeting signals of microlensing events. They combined data from four ground-based telescopes and two space telescopes to make the discovery. Hervé Bouy, a member of the team, said, "We used tens of thousands of wide-field images from ESO facilities, corresponding to hundreds of hours of observations, and literally tens of terabytes of data. There could be several billions of these free-floating planets roaming in the Milky Way without a host star."[17]

The number of rogue planets out there is still highly uncertain. Simulations suggest that the number of nomads rivals or exceeds the number of stars. The first detection of an exoplanet in another galaxy, an incredible 3.8 billion light-years away, led to an estimate of two thousand Moon to Jupiter-mass objects per main-sequence star.[18] Another study by researchers at Stanford suggested an eye-popping hundred thousand times more nomad planets than the number of stars in the Milky Way. Alan Boss, who we last met pinching himself for failing to predict planet migration, commented, "To paraphrase Dorothy from the Wizard of Oz, if correct, this extrapolation implies we are not in Kansas anymore and we were never in Kansas."[19] The universe is riddled with unseen planetary-mass objects that we're just now able to detect. Even if there are only as many nomads as stars, that would be a prodigious several hundred billion in our galaxy.

Perhaps there's one close to home. As we've just seen, Nesvorný's best match to the current configuration of our solar system came from assuming it initially had four gas giants and one ice giant. The ice giant, with a mass comparable to Uranus and Neptune, was ejected by Jupiter. It was separated at birth. It's an intriguing thought: one of our siblings might now be a rogue planet, sailing through interstellar space.

Rogue planets between the mass of Neptune and Earth could be habitable. Interior rocks create heat by radioactive decay, and a thick atmosphere acts as an insulating blanket. They could sustain liquid oceans and geological activity, while seafloor volcanism could provide the energy required for life, as probably happened on Earth.[20] They could be living planets, with self-contained ecosystems, no star required.

What kind of journey do rogue planets undertake? The escape velocity of Earth from the solar system is forty-two kilometers per second or nearly a hundred thousand mph. An interaction that kicks a planet out of a solar system might give it a speed of a hundred kilometers per second. At that speed, it would take thirty thousand years to traverse the distance to the nearest stars. Over millions of years, the rogue planet would pass hundreds of stars. Space is vast, and solar systems are tiny compared to the space between them, so it's unlikely that any rogue planet would pass close enough to be lassoed by gravity and join another planetary system. Sailing through the void, it might look longingly at other planetary families, but it would remain an orphan.

Some planets can make an even more dramatic journey. In 1988, astronomer Jack Hills predicted that a tightly bound binary star system coming close to a million-solar-mass black hole could cause one star to be bound to the black hole and the other to be ejected at four thousand kilometers per second.[21] At that speed, close to ten million mph, the star and its planets would leave the galaxy in a million years. The Milky Way has a massive black hole, and in the 1990s, several dozens of these "hypervelocity" stars were discovered. As many as a thousand could have been ejected by our galaxy's black hole—a minuscule fraction of the stars in the galaxy.[22] The fate of planets in these violent black hole encounters is interesting. Some continue to orbit their star as it is ejected from the galaxy, some get ripped from their star and devoured by the black hole, and some are ripped from their star and ejected as intergalactic rogue planets.[23]

Ejection by a massive black hole turns out to be a minor path for stars and planets to go rogue. Galaxy interactions and mergers can throw huge numbers of stars out of their galaxies. One study found that as many as half the stars in the universe live outside galaxies.[24] And when the supermassive black holes in big galaxies interact and merge, they can propel stars out of the merger at a phenomenal speed, between 10 and 30 percent of the speed of light. At those speeds, they can travel substantial distances across the universe. This may be an adventure that awaits us in three billion years when the Milky Way merges with the Andromeda galaxy since our solar system has a chance of being ejected as a result.

According to Avi Loeb, a leading astronomer who has studied hypervelocity stars and their planets, this research has a profound implication for life in the universe. Loeb is a brilliant cosmologist who has written eight hundred papers and eight books, and chaired the astronomy department at Harvard for a decade. We'll encounter him again later with his provocative argument that we've been visited by an alien artifact. He thinks hypervelocity stars and planets are a transit system for life in the universe. As he puts it, "Tightly bound planets can join the stars for the ride. The fastest stars traverse billions of light-years through the universe, offering a thrilling cosmic journey for extra-terrestrial civilizations. In the past, astronomers considered the possibility of transferring life between planets within the solar system and maybe through our Milky Way galaxy. But this newly predicted population of stars can transport life between galaxies across the entire universe."[25]

"I am tormented with an everlasting itch for things remote. I love to sail forbidden seas and land on barbarous coasts."[26] So wrote Herman Melville in *Moby-Dick*, speaking for all wanderers. How might we imagine life on a rogue planet?

Our rogue planet is born attached to a star system near the center of a spiral galaxy. Passing dangerously close to the galaxy's massive black hole, it is ripped free from its star and sent hurtling away from the black hole. This super-Earth has a molten metal core and violent volcanic activity. Strong gravity means the dense atmosphere clings like a shroud. Life starts much as it did on Earth, in nutrient-rich, boiling water near geysers. On this alien world, as on Earth, radioactive isotopes were more common after formation. Radioactivity releases energy and electrons that drive chemical reactions. Hydrogen is abundant, retained by strong gravity. A dynamic environment at the interface of rock and water breaks down inorganic compounds, and converts them to complex organic molecules. Natural processes generate fats and proteins, and fatty molecules combine to form semipermeable spheres, which are primitive precursors to a cell. A hundred million years later, the rudiments of a metabolism emerge, built around a large molecule that can store heritable information. It's not RNA, but it might be RNA's distant cousin. The planet is passing through the membrane from chemistry to biology.[27]

With no star, the rogue planet has no day-night cycle and no seasons. There's also no danger of an "end of days" as a star runs out of fuel; radioactive decay in this planet's core can sustain life for many eons. Darwin has a long reach. Natural selection plays out here as well. Random mutations lead to genetic variation, and organisms survive and proliferate based on their fitness to the environment. What comes next is pure speculation. The gloom is only occasionally broken by lightning and volcanic activity.

Sensing heat has more selective advantage than sensing light, so organisms evolve using infrared sensors. Creatures develop central nervous systems and rudimentary brains to collect as well as process information, eat, and avoid being eaten. They're small and squat to function in the thick air and strong gravity. It might not happen at all, but let's imagine that at some point, after a few hundred million or maybe billions of years—the universe has nothing but time on its hands—a subset of these strange creatures becomes sentient.

What then? Do they communicate with each other? Are they intelligent? Do they have hopes and dreams and fears? Arthur Eddington was one of the greatest astrophysicists of the twentieth century—the man who figured out how the stars shine. He was supremely confident in the power of physics to understand how the universe works. Eddington speculated that an intelligent species on a cloud-shrouded planet could deduce the laws of physics that apply throughout the universe, on rogue planets as they do on Earth. If they did, they could predict that gravity forms stars, even if they'd never seen stars.[28] One day, intelligent creatures on the rogue planet might develop technology to peer through the clouds or launch themselves into orbit. And they would see that they inhabit a "starship" sailing through a vast intergalactic void.

III
The Search for Life beyond Earth

Despite the almost certain abundance of habitable real estate in the universe, it's conceivable that life on Earth was a unique accident. The thought that we're alone in a vast universe is disquieting. We won't know if there is biology anywhere else unless we look. A thorough search means the direct exploration of planets and moons in our solar system, searching for atmospheres of earthlike exoplanets that have been altered by microbes and listening for artificial signals from technological civilizations. Time will tell if this is a quixotic quest.

Astronomers are generally optimistic that given temperate conditions and sufficient time, simple molecules naturally combine, and chemical complexity leads to replicating molecules and structures like cells that concentrate genetic information. If microbial life is common beyond our solar system, its mechanisms will inform our effort to nurture Earth, where life is also primarily microbial. The existence of intelligent life elsewhere is a long shot. If it does exist, however, it's unlikely that we're the first species to reach our level of intelligence or are the smartest kids on the block. Perhaps "they" are malign, and we take a risk by even trying to communicate. But if "they" are beneficent and wise, we could gain knowledge to help our species—and all life on Earth—endure and prosper. In the next chapters, we learn that insight into our terrestrial challenges may come from the discovery of life beyond Earth.

14
The Solar System

"It would be great news to me to find that Mars is a completely sterile planet. Dead rocks and lifeless sands would lift my spirits. On the other hand, if we discovered traces of some simple extinct life form: a bacterium, some algae, it would be bad news. If we found fossils of something even more advanced, like a trilobite or even the skeleton of a small mammal, it would be horrible news. The more complex the life we found, the more depressing the news."[1]

These are the words of Nick Bostrom, philosopher, polymath, and founder of the Future of Humanity Institute at Oxford University. Bostrom is one of the deepest thinkers about the implications of finding life beyond Earth. As we ramp up the search for life on Mars, and in our solar system more generally, why is he so gloomy about a discovery that most scientists fervently hope for?

To see why Bostrom is so worried about the philosophical consequences of finding life in our solar system, we need to consider something called Fermi's question—also called "the great silence." (I'll revisit it at the end of this part of the book.)

Enrico Fermi was a famous physicist who, in 1950, wondered why intelligent aliens had not been found or made their presence known. "Where are they?" he asked.[2] With myriad planets out there, many of them habitable, and a lot of time for intelligent species to evolve, there must be many of them. To explain their apparent absence, economics professor Robin Hanson proposed the idea of a "Great Filter."[3] According to Hanson, the Great Filter is an evolutionary hurdle that trips up most forms of life, preventing them from evolving to an advanced state in which they can travel through space and become a cosmic species. The filter could act in the early stages—making it highly unlikely for even the most basic life-forms to arise, for example. If the filter does typically act at such an early stage or any stage that we on Earth have already surpassed, then we still have a shot at longevity and extending our reach out into deep space. Finding life in our cosmic backyard—elsewhere in our solar system—would suggest that there's no filter at such an early stage. We see now why Bostrom hopes Mars is dead. If we find nematodes there or any multicellular life, it means the filter is later in the evolutionary time line. That makes it more likely that it lies ahead of us and we are facing doom. In other words, the most ominous filter would be the one that lies in our future: instability due to technology.

This argument serves as an unsettling backdrop as we consider the search for life beyond Earth. Despite this misgiving, our curiosity alone is enough to make us want to keep looking. Even Bostrom concedes that the discovery of any kind of extraterrestrial life would be "of tremendous scientific significance."[4] Those doing the search face a glaring, central question: How will we recognize "biology" on an alien world?

Two decades ago, biochemist Norman Pace made the case that biochemistry is universal. Pace is an influential biochemist and MacArthur Fellow at the University of Colorado. He noted that the elements and compounds needed to start and maintain life on Earth are widely distributed through space, but he also argued that the modes of information storage and energy utilization would be the same for biology here and elsewhere. He made a prediction: "If we go to Mars or Europa and find living creatures there and read their rRNA genes, we should not be surprised if the sequences fall into our own relatedness group, as articulated in the tree of life."[5] Life's origin on Earth involved universal chemical processes and physical conditions that apply to other planets and moons.[6]

But even if the chemical ingredients and biochemical building blocks are similar, the higher levels of organization might be radically different. We shouldn't expect to find ants or whales or mushrooms or poppies on another planet. In a homely analogy, a set of kitchens might all have ovens and cupboards with similar ingredients, but the meals that get served could vary wildly.

Within our solar system, biology at different locations need not have independent originations; it might have originated in one place and traveled through space, seeding life in other locations—an idea called "panspermia." If this were the case, it would reduce Bostrom's disappointment in the discovery of life on Mars since it would not mean that life had originated independently twice in our solar system. That would keep alive the notion that the Great Filter exists at an early developmental stage and therefore that we're among the lucky few to have got through unscathed.

If panspermia sounds like pure fiction, consider this: the debris left over from the formation of our solar system causes impacts, and impacts jettison rocks that travel through space, which gravity can then cause to fall onto another object. The solar system is like a "highway" that inefficiently carries rocks from one place to another.[7] Over a hundred meteorites have been identified as originating from Mars, and hundreds of other meteorites came here from the Moon—and a few even from Mercury.[8] Earth rocks have gone to those places as well. The transport mechanism works better for rocks that originate on lower-mass objects with thinner atmospheres, and better for rocks traveling inward, toward the Sun's strong gravity. Traffic along the interplanetary highway was heavier in the early phase of the solar system when debris was abundant and impacts common. Perhaps not by coincidence, that's when life on Earth started. Research has shown that microbes can survive the vacuum and radiation of space for years.[9] Life may have originated independently on Earth and Mars. Or it may have started here and got to Mars aboard meteorites, seeding life there. Or it may have started on Mars and come here on a spray of Martian meteorites. The transfer of rocks containing viable organisms from Mars to Earth is a hundred times more frequent than the transfer from Earth to Mars.[10] That makes it possible that we're all Martians.

For now, we can only look for life as we know it. If biology expresses in radically different ways off Earth, we don't know how to build an experiment to detect it. Biologists have begun to tinker with the tool kit of life in a field called synthetic biology. They can use unnatural molecules to reproduce behavior from natural biology, with the goal of creating artificial life.[11] But it's much more difficult to jump completely out of the box and wonder, "How strange could life be?"

Later I'll consider the search for intelligent life and advanced civilizations across the galaxy. We might have some confidence that microbial life everywhere shares common attributes, but we're reduced to pure speculation regarding extraterrestrial intelligence and technology. Let's start our journey of a thousand miles with a single step, looking in our backyard at our nearest, marginally habitable neighbor.

Mars has been the object of our attention and sometimes obsession for centuries. By the early nineteenth century, astronomers knew that Mars had a day similar in length to Earth's day, seasons, ice caps that changed with the seasons, and surface markings suggesting vegetation. Fictional accounts of life of Mars began to appear in print, and late in the century, scientists were forced to respond to the bold claims of Percival Lowell. Lowell was a wealthy Boston merchant with a keen interest in astronomy. After he retired, he traveled west and built a telescope to take advantage of the pristine desert skies in northern Arizona. As Mars made its closest approach to Earth in a decade, he described linear features on the surface and became convinced they were canals built to carry water from the poles to a dying civilization living near the equator. Lowell publicized his findings in a series of popular books.[12] He was fooled by a combination of an optical illusion and wishful thinking, but his idea took root in the popular imagination.[13]

I remember the first time I saw a pitch-black night sky. As a graduate student, I left the leaden gray skies of my native Edinburgh to visit Steward Observatory, where I now work. On Mount Lemmon, at an elevation of three thousand meters, the stars shine like jewels set on black velvet. The stars twinkle due to turbulent air motions in the upper atmosphere, but when this turbulence is mild, they shrink to pinpoints of light. I pointed my binoculars at the Moon and the view was breathtakingly clear. When the turbulence abated and the "seeing" became good, the general blur of geological features sharpened. My imagination took over, and it was easy to see animals, faces, and linear features. I could sympathize with Lowell, who three hundred miles away and over a century ago, had imagined signs of life on the Red Planet.

Three years after Lowell published his findings, H. G. Wells was inspired to write a book that has never been out of print. *The War of the Worlds* starts with a chilling description of inhabitants on Mars who are plotting to invade Earth: "Yet across the gulf of space, minds that are to our minds as ours are to those of the beasts that perish, intellects vast and cool and unsympathetic, regarded this earth with envious eyes, and slowly and surely drew their plans against us."[14]

The existence of life on Mars was unquestioned through the first half of the twentieth century. Edgar Rice Burroughs, better known for his Tarzan adventures, wrote a series of books set on Mars starring the heroic John Carter. This is classic pulp fiction, and it features a Mars inhabited by fierce warriors, buxom princesses, and exotic animals. We can see an echo of Lowell's vision of Mars in Burrough's fantasies as well as in the classic science fiction of the mid-twentieth century, such as in *Red Planet* by Robert Heinlein and *The Martian Chronicles* by Ray Bradbury. *The War of the Worlds* inspired half a dozen films, various comic books, several TV series, and numerous stories by other authors. Most infamously, it was dramatized in a 1938 radio show by the actor and future film director Orson Welles. Using the format of a live music performance interrupted by news bulletins, the show featured increasingly alarming updates about a Martian invasion and the futile efforts of the military to stop it. With the storm clouds of war building in Europe, the United States was an anxious country, so it's not surprising the broadcast caused local panic.[15] It cemented Welles's reputation as a brilliant provocateur.

The hopes that Mars was a living planet were dashed in 1965 when the orbiting space probe Mariner 4 returned the first close-up pictures. Spectroscopy using ground-based telescopes had already shown that Mars has a thin atmosphere with almost no water vapor, but Mariner 4 saw no oceans, no rivers, and no signs of life. The surface was covered with craters, indicating no plate tectonics or weathering of any kind for billions of years. Mars had no magnetic field to protect against harmful cosmic rays. It was a barren, frigid desert.[16]

Since then, the pendulum has swung back toward a view of Mars as habitable—just not with intelligent, malevolent creatures. We know a lot because there have been nearly fifty missions there since 1960. The failure rate was 80 percent in the 1960s, but that has decreased to 10 percent since 2000, even as the missions became more complex and challenging.[17]

Orbiters showed features that on Earth are associated with water, such as shorelines, river deltas, and gullies running downhill. Mars was almost certainly habitable until three billion years ago. There was sufficient water back then to make a global layer a kilometer deep. But with weak gravity and no magnetic field to protect against a battering by cosmic rays, Mars couldn't hang onto its atmosphere. It slowly evaporated into space, and the surface water was steadily sequestered into the crust.[18] As explained in chapter 12, water can exist underground, kept liquid by the pressure of overlying rock and heating from radioactive decay within the rock. There are likely to be subsurface aquifers all over Mars, and radar has been used to locate underground lakes in the south polar region.[19] Indirect evidence indicates current life on Mars.[20] This includes seasonally varying methane—which on Earth is due to biological activity—and also results produced by one of the 1970s' experiments onboard the Viking lander probes, whose designer insists pointed to biology.[21] Indirect evidence is not good enough to sway the skeptics, however. Unequivocal evidence of past or present life on Mars is elusive, so the planet continues to be the subject of intense scrutiny.

Over the years, Mars rovers have increased in size from the equivalent of a shoebox to a go-kart to an SUV. Despite the sophistication of Curiosity and its beefier cousin Perseverance, none of the rovers have had life-detection experiments. The ExoMars mission will deliver a rover to Mars in 2023. The rover will drill down two meters and analyze subsurface samples, but although it was named for the DNA pioneer Rosalind Franklin, its Life Marker Chip experiment is no longer part of the payload. Perseverance scooped up its first Mars sample in 2021.[22] It also detected organic materials in the Jezero crater in 2021.[23] To be clear, this isn't a detection of Mars life. But it's exciting to find the building blocks for life in a place that was covered with water in the past. Perseverance will stash forty rock samples, but returning them to Earth will take at least ten years.[24]

This rate of progress is frustrating to astrobiologists, so they practice by finding patches of Mars on Earth. Chris McKay has been doing fieldwork in remote parts of the world for thirty years as a senior scientist at NASA Ames Research Center. "Deserts are particularly relevant for the study of life on Mars because Mars is a desert world," he explains. "When we study deserts on Earth, we see many of the chemical and biological processes that we think happen on Mars. We see oxidants in the soil, we see challenges in the preservation of organic material, we see life trying to adapt at very low levels of moisture. These are all themes that keep occurring when we think about life on Mars."[25]

The best proxies for Mars are the dry valleys of Antarctica and the Atacama Desert. "The Atacama Desert is the only place on Earth where Viking could have landed, scooped up soil, and failed to find evidence of life. It would have got the same results: no organics, no life, but the presence of some kind of chemical reactions in the soil. Yet walk or drive a hundred kilometers south and there are a million bacteria in a gram of soil." This fact drives home the importance of choosing the best location from which to take a sample for return. McKay has no hesitation. "If the NASA Administrator said, 'Here's a couple of billion dollars; do what you think is the best thing to do on Mars,' I would send a mission to the south polar region—in fact, to the crash site of the Mars Polar Lander, 76 degrees south, in that ancient ice-ridge and crater terrain with the crustal magnetic features. I would send a sterilized deep drill to go down into that ancient ice and bring back samples of the ancient permafrost material. Then search it, not just for fossils, but for actual preserved, frozen, dead Martian life forms."[26]

Remote fieldwork is a pleasure for McKay. "The most fun days I remember were days out in the field, out in the Mojave or in the Atacama or even the Antarctic, even though it's minus 20 degrees and we're freezing cold. There we do things that people in other jobs pay to go do." He laughs as he remembers putting in a travel request to John Rummel at NASA Headquarters. "I said I'm going to the desert in Namibia in Africa. And he wrote back 'is this a vacation or is it work?' I didn't give him my honest answer. I said 'John, we're sleeping in the dirt, we're eating dirt, we're digging in the dirt. If you want to call that a vacation, you can. I call it work.' So, he approved the travel."[27]

Like the spoiled child of doting parents, Mars seems to get all the attention. But astrobiologists insist there are a dozen or more

Figure 14.1

Journey to Mars. NASA's plans for Mars exploration involve a coordinated set of landers, orbiters, and eventually, life-detection experiments and sample return missions. The launch vehicles are being built by a combination of NASA and commercial space companies like SpaceX. (Credit: NASA, "NASA's Journey to Mars," last updated 2017, https://www.nasa.gov/content/nasas-journey-to-mars.)

locations in our solar system with the ingredients for life. I consider them next.

Before we turn our gaze to the icy moons of the outer solar system let's glance back briefly at Venus. Earth's twin is at the inner edge of the habitable zone, but vigorous volcanic activity billions of years ago led to a runaway greenhouse effect. The planet named for the goddess of love has a surface temperature of 460°C (860°F) and a smothering carbon dioxide atmosphere nearly a hundred times denser than Earth's. Venus seems inimical to life, but there's a temperate zone at midlevels in the atmosphere. A controversial detection of phosphine was claimed in 2020. Phosphine on Earth is always associated with biology, and there are also unexplained chemical imbalances in the atmosphere that might indicate biological activity.[28] Further analysis of the phosphine detection in 2021 showed that the signal probably came from sulfur dioxide, but that doesn't yet mean that life on Venus is completely ruled out.

The discovery of things living in Venus's clouds wouldn't surprise David Grinspoon. He is a senior scientist at the Planetary Science Institute, an accomplished guitarist, and the winner of a PEN award for nonfiction writing. About the clouds, he says, "They're within the right temperature range for life as we know it, and they're in a continuous dynamic environment, one with lots of interesting energy sources and chemical equilibrium that has not yet been well explained. It is an aqueous environment, albeit one suffused with concentrated sulfuric acid."[29] Venus hasn't featured prominently in NASA's plans for decades, but in June 2021, the agency awarded two $500 million development grants to Discovery Program missions to Venus. DAVINCI+ will send a descent sphere to plunge through the thick atmosphere, mapping it and determining if the planet ever had an ocean. VERITAS will use radar to map the surface and help deduce the planet's geological history. These probes will reach Venus at the end of this decade.[30] A multi-billion-dollar flagship is also vying for funding.[31] It would send three orbiters, a lander, and a balloon to float in the upper atmosphere. If it gets the nod, we'll have to wait until the mid-2030s before Venus reveals all of its secrets.

Most of the attention of astrobiologists is taken by objects much farther from the Sun than Venus. It takes patience and persistence to explore the outer solar system. NASA gets dozens of proposals for each mission category; a team might work for a decade on a concept only to get booted at the last stage of the competition. Then it's a decade for development, and a decade more to reach the destination. Carolyn Porco compares the process to the peaks and valleys of raising a child. She led the imaging team on the Cassini mission to Saturn. Porco says, "It's similar in terms of the level of commitment. There are joyous moments, but also outrageous frustrations, and lots of hand wringing. This mission required inordinately long periods of time when I had to be obsessively devoted to it."[32] This type of parenting is subject to the election cycle and whims of the US Congress too.

ESA is launching the Jupiter Icy Moons Explorer in 2023. The spacecraft will visit Ganymede, Callisto, and Europa—all with subsurface oceans (see chapter 12)—after it arrives at Jupiter in 2030.[33] For its part, NASA is launching Europa Clipper in 2024. The Europa Clipper will sniff Europa's thin atmosphere, map the ice shell's thickness, and sample some of the plumes that rise from the surface. A subsequent lander to search for life is still under consideration. We can visualize it being lowered to the surface by a UFO-like sky crane, such as those that delivered Curiosity and Perseverance to Mars. Near the frozen ground, an autonomous navigation system takes over as it tries to find a smooth patch between the crevasses and tall ice spikes.

NASA and ESA planned a joint mission to explore Saturn and its moons Titan and Enceladus, but it was shelved to fund the Jupiter missions. Recognizing that it needed to become nimbler, NASA developed the innovative Dragonfly mission to Titan. A quadcopter drone will cover hundreds of kilometers, sampling the chemistry of the surface and the methane-ethane lakes. Cryovolcanoes mix water into the organic soup, and feeble yet reliable sunlight drives the moon's photochemistry.[34] Dragonfly is scheduled to launch in 2027. Elizabeth Turtle, the mission's leader, says, "We have all these ingredients necessary for life as we know it, and they're sitting there doing chemistry experiments on the surface of Titan. That's why we want to send a lander there."[35] If life exists on Titan, it will have a different biochemical basis for life than on Earth and so truly would be "Life 2.0."

Tiny Enceladus is also a compelling target. NASA scientists have a strategy for sample return from the surface, but it has a price tag of billions of dollars and a time frame of two decades.[36] Not to be outdone, Russian billionaire Yuri Milner is proposing to send a probe to search for microbes in the plumes of ice particles ejected from the subsurface ocean of Enceladus. The Breakthrough Enceladus mission would be the first privately funded deep space probe, and it would get to its target twice as fast as a NASA mission and at a tenth of the cost.[37]

Even more ambitious dreamers plan to visit Neptune and its moons as well as Pluto. Such missions would not return data until 2050. We still know so little about these remote regions that we're in the exploration phase. Detailed studies and life detection shimmer in the distance. But with persistence along with a few billion dollars here and there, if there's life in our solar system, we will find it in the coming decades.

15

Sniffing Biosignatures

The next frontier of astrobiology is the search for indications of life on the most earthlike of the thousands of exoplanets that have been discovered. Even the closest of those planets are hundreds of trillions of miles away. Astronomers can't hope to produce images of these planets and so they must use more indirect methods to find evidence that these distant worlds potentially harbor life.

To put this into perspective, I begin this chapter by looking the other way through the telescope and considering how we would detect life on Earth from a remote vantage point. It's not as easy as you might think.

Seen from a trillion miles away, Earth would be a single pixel in an image, a pale blue dot. None of the great construction feats of human civilization—the pyramids of Egypt or Great Wall of China—would be visible. We've built enormous cities, and although 50 percent of the world's population now lives in towns or cities, less than 1 percent of Earth's surface is urbanized. Even the harm we've done to the planet through fires and deforestation and desertification wouldn't be detectable. Since the beginning of the industrial age, we've increased the amount of carbon dioxide in the atmosphere by 30 percent, but carbon dioxide is a trace ingredient of the atmosphere, so that impact on the overall chemical composition is a tiny 0.01 percent.

As we'll see later, imaging techniques for exoplanets are improving. With persistent observations over months and years, spectra of exoplanets might show daily variations due to their rotation and seasonal variations if there's an orbital tilt. A group at Caltech used an Earth observation satellite to test out this idea. With ten thousand images of our planet taken over two years, each degraded to a single pixel, the researchers measured two different sets of changes: those caused by varying cloud cover across the planet, and others caused by continents and oceans rotating into view.[1] It was an exciting proof of concept for exoplanets, but still topography, not biology.

In any spectral analysis of Earth's atmosphere, however, one chemical component would stick out like a sore thumb: oxygen.

This was the thought that occurred to James Lovelock after he was hired by NASA in 1961 to plan for the detection of life on Mars. Lovelock wasn't an obvious choice to help design a planetary probe. His PhD was in medicine, and his early claim to fame had been a sensitive tool for measuring gas concentrations in the lab. In the late 1950s, his tool was used outside the lab to map pesticides and environmental contaminants for the first time, inspiring Rachel Carson to write *Silent Spring*, which was published in 1962. Carson's book launched the modern environmental movement.

Hired by NASA to work at the Jet Propulsion Lab in Southern California, Lovelock brought his outsider's perspective to the team making instruments for Mars. He recalls, "I found that what they had in mind was almost unbelievably naive. They seemed to assume that because Mars was a desert, they'd be looking for the same sort of life they'd find in the Mojave Desert. My natural question was: How on earth do you know, if there is life on Mars, it's anything like here?" To Lovelock, the key was entropy or disorder. "Inanimate things tend to end up in equilibrium sooner or later, and that is a state of very high entropy, or low order—whereas something alive, which is infinitely more complex, will have a very low entropy indeed. Anyway, I got called in to see the director of the lab and he said, 'You've got until Friday to give me an experiment we can send to Mars.' I went away and thought about it, and by Friday, I could tell him: It's really quite simple. All you have to do is analyze the atmosphere of Mars. If there's life on the planet it's bound to get its raw materials from the atmosphere, as there isn't any other mobile medium— there's no ocean—and it would dump its waste products in the atmosphere for the same reason. And those two processes would change the composition of the atmosphere in a way that would give it less equilibrium than it would have on a dead planet."[2]

Lovelock looked at the atmospheres of Venus, Earth, and Mars, and deduced that the presence of life on any planet would be reflected in the atmosphere of the planet. The differences are simple but startling. The atmospheres of Venus and Mars each contain 95 percent carbon dioxide with trace amounts of oxygen. By contrast, the atmosphere of Earth contains 21 percent oxygen with a trace amount of carbon

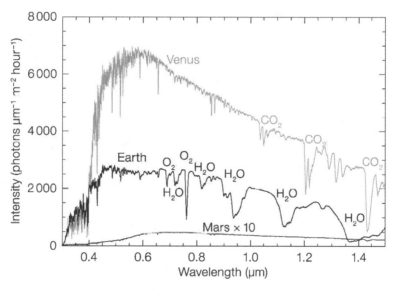

Figure 15.1

Spectra of Venus, Mars, and Earth at optical and near-infrared wavelengths. Venus and Mars have strong signatures of carbon dioxide, while Earth has spectral features due to oxygen and water vapor, indicating life on the planet. (Credit: ESO, adapted from "How to Characterize Habitable Worlds and Signs of Life" by L. Kaltenegger, 2017, *Annual Review of Astronomy and Astrophysics*, volume 55, https://arxiv.org/ftp /arxiv/papers/1911/1911.05597.pdf, 24.)

dioxide. Cyanobacteria first oxygenated Earth's atmosphere billions of years ago, and oxygen has made up a significant portion of the air ever since, maintained by cyanobacteria and plants.[3]

Imagine the biosphere shuts down overnight. All life on Earth is extinguished. The only way a distant observer would detect the catastrophe is by changes in the composition of Earth's atmosphere. Without plants and cyanobacteria to sustain it, all the oxygen would disappear in about ten million years—the blink of an eye in geological time—as it dissolved in seawater, reacted with rocks, and rusted metals. If we reverse this logic, the presence of the unstable and reactive gas oxygen in a planet's atmosphere is a telltale sign of life.

Lovelock had a straightforward answer for NASA. The atmosphere of Mars is close to chemical equilibrium: nearly all carbon dioxide and little else. He suggested they save their money and not bother sending astronauts there. Mars is probably dead. But he was left with a puzzle about Earth. "The more I looked at our atmosphere, the more astonishing it was. It was weird, weird beyond belief. You have things like oxygen and methane co-existing, for one thing. The chances of this occurring by ordinary, random chemistry are so remote that no computer will even give you a figure. Then there was far too little carbon dioxide. It's pouring out of volcanoes all the time, so where the heck is it all going? I mean, it's a trace gas on Earth—it was only about 300 parts per million at that time. Then there were gases like nitrous oxide that can only be made biologically. When you looked at the whole atmosphere, you found that everything either was made or had been profoundly modified in amount by life at the surface."[4] It included nitrogen, which is four-fifths of the atmosphere. If Earth was dead, all that nitrogen would end up in the sea. It's only returned to the atmosphere because bacteria break nitrates down and release nitrogen gas.

Lovelock was talking to Sagan when the insight came to him in a flash: Earth's atmosphere must be regulated by life. Sagan initially pushed back, but he knew that the Sun has warmed up by 30 percent since life on Earth started, and one of the big puzzles in astronomy is: Why has Earth's temperature stayed constant? Lovelock called the idea Gaia, with the name suggested to him by novelist William Golding, who lived in the same small village in Devon. He developed the theory with Lynn Margulis, the evolutionary biologist who first pointed to the importance of symbiosis in evolution. Margulis said, "Life does not exist *on* Earth's surface so much as it *is* Earth's surface. . . . Earth is no more a planet-sized chunk of rock inhabited with life than your body is a skeleton infested with cells."[5] Gaia was a radical challenge to traditional biology. In Darwin's view, life was passive, forced to adapt to a particular environment. Gaia says that life and the environment are one coevolutionary process. Life acts as a planetary thermostat.[6] Without life, carbon dioxide would flood the atmosphere and Earth would be hot enough for the oceans to boil. Life has kept the oxygen steady for three hundred million years. A few percentage points higher and forest fires would engulf the planet. But a few points lower and all animals would perish.

Gaia is misunderstood. Lovelock patiently played down its New Age trappings. Earth is not a superorganism. To those who say it smacks of teleology or design, he noted that the theory makes testable predictions, and many have been borne out.[7] Lovelock was an iconoclast, working independently for most of his professional life from a barn that he converted into a laboratory in a rural corner of southwest England. He died in 2022, but continued writing books and issuing darkly humorous warnings about the damage we're doing to the planet after he turned one hundred.

A biosignature gas is any gas that is created by life that accumulates to a detectable level in a planetary atmosphere. In addition to oxygen, primary targets are ozone, methane, nitrous oxide, and water vapor. I emphasized oxygen because of its direct connection with biology and high concentration in the modern atmosphere. Yet as we've seen in the work of Lisa Kaltenegger, the composition of the atmosphere has changed dramatically since Earth formed.

There was virtually no oxygen during the Archean eon, from 4 to 2.5 billion years ago. With the emergence of blue-green bacteria and photosynthesis, carbon dioxide became life-giving food. Oxygen was produced as a waste product, at a low enough level to dissolve in the oceans and stay at a low level in the atmosphere. Around 2 billion years ago, the process gave rise to a planetary crisis, the "Oxygen Holocaust," because oxygen was highly toxic to the first bacteria and killed them off. The crisis provided a window of opportunity for new types of organisms such as algae and plants. By a billion years ago, all the "sinks" for oxygen in the sea and on the land were filled, and it began its rise to the present-day levels.[8]

Oxygen is an excellent biosignature, but Earth had long periods when there was life yet oxygen was at a low level. Also, there are scenarios where oxygen can be produced by geology on a lifeless planet, resulting in a "false positive" detection.[9] In the first few billion years after life started, when there was little oxygen, the combination of methane and carbon dioxide is a potential biosignature.[10] No single gas is a smoking gun, so a spectrum with a set of biomarkers needs to be compared to models with and without biology.

Vikki Meadows is the guru that many astrobiologists turn to as they prepare for the data on exoplanet atmospheres. She runs NASA's Virtual Planet Laboratory. Meadows sounds like a sorcerer's apprentice, but her team does not actually make planets, real or imagined. The team simulates a full range of plausible atmospheric and geological properties in a self-consistent way so that biomarkers can be correctly interpreted. Meadows was born in Australia, but left after her PhD dissertation to work at Caltech. Her planets are virtual and so is her collaboration, with researchers at eighteen institutions around the world. The key to characterizing a planet and looking for life is a spectrum. "We break reflected light from the planet into as many constituent wavelengths as we can. There's a tradeoff between the number of photons we're able to collect in a given channel versus how much detail we would like to see in the spectrum of the planet. When we get the spectrum, it's going to be from the entire face of the planet. All the continents, oceans, clouds—everything's crushed together into a disk average. Even so, it's extremely powerful, because within that spectrum will be the signatures of atmospheric composition, and perhaps what the surface is like, whether or not there's vegetation, and whether or not there's life."[11]

Detecting spectral biosignatures in exoplanet atmospheres is a challenge. The feeble reflected light from the planet, typically one-billionth as bright as the light from the host star, must be spread into a spectrum. As described in chapter 4, the experiment gets easier by observing in the infrared and picking exoplanets that orbit dim red dwarfs. Targets will mostly be transiting exoplanets since they allow the atmosphere to be backlit by the star and thus spectroscopy can sniff the chemical composition. Here's how life might be detected in the next five to ten years.

Sara Seager will be our guide. She's devoted her career with a laser-like focus to the ways exoplanets can be characterized with the next generation of ground- and space-based telescopes. Seager made an exhaustive list of fourteen thousand stable and volatile molecules that might be biogenic, including many that might arise in alternative biochemistries.[12] The first spectra obtained for earth-like planets will be noisy and of frustrating quality. Just as no single molecule is an unequivocal indicator of biology, no single spectrum will be convincing proof of life elsewhere. But the data will accumulate, and Seager notes, "Although we may not be able to point to a planet with certainty and say, 'that planet has signs of life,' with enough rocky worlds with biosignature gases, we will inspire confidence that life not only exists in the solar neighborhood but is common in our Galaxy."[13] Seager is familiar with pushing through obstacles. When her husband died of cancer, the bottom dropped out of her world, and she almost quit. Obsessed by her quest, she never shopped for groceries or cooked or pumped gas. All she had to do was find another Earth. "I had worked so hard," she says. "I had all the years I called the lost years with Mike when I ignored him. We had little kids. I was working all the time, exhausted all the time."[14] Now she is remarried and has restored balance in her life. She keenly anticipates new tools in the search for biosignatures. First up is the James Webb Space Telescope, launched in December 2021.

The JWST is big in every sense. It's the largest telescope ever placed in space—a hundred times more powerful than the venerable Hubble Space Telescope. Its orbit takes it 1.6 million kilometers (1 million miles) above Earth's surface, far beyond Hubble's 540-kilometer (340-mile) altitude. It has a gold-plated mirror so large that it had to fold like origami to fit into the rocket and then it unfolded like a "transformer" out in space. Since it will work primarily in the infrared, the telescope has a five-layer sun shield giving it a Sun protection factor of a million and keeping the telescope at −230°C (−380°F). The budget is massive too. First planned for launch in 2007 with a budget of $500 million, the JWST launched fourteen years late with a budget that ballooned over $11 billion. Dubbed "too big to fail" and the "telescope that ate astronomy," the facility cannot be fixed by astronauts with wrenches as Hubble was five times.[15] After a nerve-wracking but successful launch and deployment of the mirrors and sun shield, JWST successfully delivered its first science data after the six-month commissioning phase. Astronomers anticipate receiving unique and thought-provoking observations.

The killer app for the JWST is detecting "first light" in the universe: the first wave of stars and galaxies that formed soon after the big bang.[16] The telescope wasn't designed to study exoplanets, so it will be limited to several dozen sub-Neptunes and a few super-Earths. Infrared spectra will focus on water, methane, and carbon dioxide as biosignatures, rather than oxygen. For example, Hycean planets, the large water worlds discussed in chapter 10, will make compelling targets in the search for life. Habitable planets in the TRAPPIST-1 system will also be among the JWST's first targets.[17]

Meanwhile, a new generation of giant ground-based telescopes will start gathering data in the late 2020s. They are the 24.5-meter (80-foot) Giant Magellan Telescope in Chile, the Thirty Meter Telescope in Hawaii (30 meters is 100 feet), and the 39-meter (128-foot) Extremely Large Telescope, also in Chile. All will have high-resolution spectrographs that are sensitive to oxygen and a wide array of other biosignatures. They might even have the sensitivity to detect the "red edge" that's a spectral signature of chlorophyll and the plants covering 60 percent of Earth's surface.[18] These telescopes will have coronagraphs to blot out the host star's light (see chapter 4) and enough light-gathering power to make images of pale blue dots tens of light-years away.[19] In space, ESA will launch PLATO, which will feed candidates for characterization to large telescopes on the ground, and ARIEL, which aims for transit spectroscopy for a thousand exoplanets. ARIEL has a smaller mirror than the JWST, but its entire mission is devoted to exoplanets.[20] From NASA, the Nancy Grace Roman Space Telescope is expected to launch by the end of the decade. As we learned earlier, Seager leads the effort to add a free-flying starshade, which would provide another tool for characterizing exo-Earths.

Looking forward to the 2030s, budgets are uncertain and the crystal ball is cloudy. NASA is considering HabEx (Habitable Planet Imaging Mission), which would search the habitable zones of forty nearby sunlike stars, with spectroscopic follow-up for a few terrestrial planets.[21] The agency is also looking at LUVOIR (Large Ultraviolet Optical and Infrared Surveyor), with a larger mirror, access to five hundred habitable zones, and a yield of fifty terrestrial planets. One way to detect an ocean is to see the way sunlight glints off it since water is more reflective than landmass. Starlight reflecting at a shallow angle from an exoplanet reflects more light from an ocean than from a continent, and as the exoplanet rotates, the light signal changes, giving the signature of a watery world. LUVOIR has this exciting capability.

Future facilities in astronomy are prioritized every ten years through Decadal Surveys, organized by the National Academy of Sciences. Getting all US astronomers on the same page is challenging, but when they speak with one voice, it increases the odds that the projects will get funded. I've been involved in these deliberations, and it's arduous work to judge competing interests—radio waves versus X-rays—and the types of facility—space telescope versus ground-based interferometer. Fueled by endless coffee and a desire to reach consensus, we spend hours in cramped conference rooms in Washington, DC. Our eyes are always bigger than our stomachs, but the wish list must be realistic. Sometimes events move slower than the planning exercise. The JWST was a high priority two Decadal Surveys ago, but as the most recent survey was published, it hadn't yet launched. "Pathways to Habitable Worlds" is one of the top three priorities for the coming decade.[22]

There's even an audacious plan to capture images of an exoplanet that would reveal features as small as ten kilometers across. It involves using the Sun's gravity to act as an enormous magnifying glass. NASA would send a fleet of spacecraft powered by solar sails a hundred billion kilometers from Earth or sixteen times the distance to Pluto. That's the focal plane of an exoplanet a hundred light-years away being imaged by the Sun acting as a lens. Small spacecraft would dart around the focal plane gathering data that can then be assembled into a crisp and detailed image of the exoplanet. The project will take decades to realize, but NASA is excited enough about the concept to fund its further development with $2 million from its Innovative Advanced Concepts program.[23]

Unsurprisingly, Seager is out front in pushing for these new missions. In testimony to the US Congress a few years ago, she said, "To be certain of finding a large enough pool of exoplanets to search for biosignature gases, we require the ability to directly image exoplanets orbiting 1,000 or 10,000 of the nearest Sun-like stars. This will require a next-generation space telescope beyond JWST, a visible wavelength telescope with a large diameter likely exceeding 10 meters. In July 2010, I became a citizen of the United States of America, motivated by our nation's unique combination of technological forte, allocated resources for space missions, and ambitious spirit. It is within the power of our influence to cross the great historical threshold and be the first generation in human history to map the nearby exoplanetary systems and find signs of life on other Earth-like worlds."[24]

16
Send in the Nanobots

Peering into the further future, might we ever be able to make real images of earthlike planets? The challenge is daunting, and there are no telescopes with enough power, even in principle, to resolve details on an exoplanet.

There are, however, real practical solutions that involve getting much closer to exoplanets—close enough to inspect them and search for life directly. For now, that won't involve sending a human expedition. Thanks to the rapid advances in miniaturization, we can send teams of tiny robots.

I've already explored the fact that the angular separation of Earth from the Sun at a distance of thirty-three light-years is 0.1 arc seconds (see chapter 4). That's close to the limit of imaging with ground-based telescopes—the distance between millimeter tick marks on a ruler a kilometer away. Earth's diameter is one ten-thousandth the distance from Earth to the Sun, so the angular size of an Earth that far away would be 10 micro–arc seconds. Big telescopes make sharper images, but as we saw in chapter 4, there's a limit to the resolution that any telescope can reach. The method that astronomers can use to make even sharper images is called "optical interferometry," where light from separate telescopes is combined to give an angular resolution as good as a telescope as large as the separation between the telescopes.[1] The best optical interferometer at the Very Large Telescope in Chile has a size or baseline of 130 meters (427 feet) and an angular resolution of 0.002 arc seconds—two hundred times the angular extent of the distant Earth in the example above, and inadequate to produce any meaningful, detailed images.

Astronomers dream of interferometers in the stable environment of space. Design studies have been done, but there are no concrete plans. A free-flying space interferometer with a baseline of several hundred kilometers could make a ten-by-ten pixel image of our Earth analog, allowing the mapping of clouds, oceans, and continents. Those arguing for such a capability know it is far-off: "Within a period of approximately 200 years, it can reasonably be expected that high-resolution spectroscopy and then high-angular resolution direct imaging will improve considerably our knowledge of nearby exoplanets and possible global biomarkers."[2]

Two centuries is a long time to wait. Instead of remote sensing, where light is the messenger, why can't we take the direct approach and go there?

Before contemplating a trip to the stars, let's look at our modest attempts so far to leave our planet and solar system. Shrink Earth to the size of a golf ball. The Moon would be a marble held at arm's length, the Sun would be a 3-meter globe 100 meters away, and Neptune would be a tennis ball about 2 kilometers away. On this scale, the nearest star, Proxima Centauri, is fifty thousand kilometers away—an enormous distance even in a shrunken scale model.

This puts our space program into perspective. About six hundred and fifty people have experienced zero gravity in Earth orbit, but that's a modest few hundred miles from Earth, like half a day's drive straight up. Twenty-four astronauts have been to the Moon, and just twelve of those set foot on another world. That momentous achievement cost the equivalent in today's money of $280 billion, and it was so difficult and expensive that we haven't been back for over fifty years.[3] We've sent spacecraft to all the planets and many major moons in our solar system, but only five spacecraft have crossed the orbits of Neptune and Pluto. In order of their dates of launch, they are Pioneer 10 and Pioneer 11—neither of which is still functioning or transmitting signals—Voyager 2 and Voyager 1, and New Horizons. The fastest and most distant of these is Voyager 1, traveling at 17 kilometers per second (38,000 mph) and currently over 160 AU from the Sun.[4] That seems impressive, but at that rate, it would take Voyager 1 over seventy thousand years to reach the distance of the nearest star.

To send humans beyond the solar system, we're caught between a rock and a hard place. The hard place is the speed of our fastest spacecraft and the vast gulf of space to the nearest star. Tens of millennia overshadow a human life, so the only way we can make the trip is through suspended animation. The first attempts to put people "on ice" are coming not from NASA but rather from emergency room doctors trying to save patients who experience acute trauma. The patient is rapidly cooled to 10°C (50°F), and their blood is replaced with ice-cold saline.[5] The method has only been used for a few hours, with a success rate that hasn't been revealed. It's a far cry from what we'll need to reach Proxima Centauri. The sheer distance and speed needed is the hard place; the rock is a law of physics. To reach our nearest star in fifty years means traveling at 10 percent of the speed of light, which is a thousand times faster than current space probes. Due to the velocity squared term in the formula for kinetic energy, millions of times more energy is needed. Plugging in the numbers, the energy required for a one-ton spacecraft is 10^{18} joules or 130 terawatt-hours. That's the annual energy use of a hundred million US households. If this is a conventional rocket-powered spacecraft, it would also have to carry the fuel that delivers that energy—and that would need to be accelerated too. Therefore much more energy again is required. It's not going to happen.

Breakthrough Starshot has a plan that, as you will see, avoids the need to carry any fuel at all. This visionary project was announced in 2016 by Russian philanthropist Yuri Milner, Facebook founder Mark Zuckerberg, and the late Stephen Hawking. The target is the nearest star, Proxima Centauri, where there's an Earth-size planet in the habitable zone of that dim dwarf star. The distance is daunting: 40 trillion kilometers (25 trillion miles). But this project aims to send a fleet of a thousand tiny spacecraft or "nanobots" there within a generation.

The project website lays out the motivation and goal: "The story of humanity is a story of great leaps—out of Africa, across oceans, to the skies, and into space. Since Apollo 11's 'moonshot,' we have been sending our machines ahead of us—to planets, comets, even interstellar space. In the last decade and a half, rapid technological advances have opened the possibility of light-powered space travel at a significant fraction of light speed. Breakthrough Starshot aims to demonstrate proof of concept for ultra-fast light-driven nanocrafts, and lay the foundations for a first launch to Proxima Centauri within the next generation. There are three chances to find habitable planets since Alpha Centauri A and B can be visited with the same mission. Along the way, the project could generate important supplementary benefits to astronomy, including solar system exploration and detection of Earth-crossing asteroids."[6]

The road map for the project was laid out by University of California physicist Phil Lubin, who is on the project's high-powered Management and Advisory Committee.[7] Interstellar travel sounds incredibly ambitious or almost fanciful. How is it going to work?

For propulsion, it's a matter of elimination. The entire history of rocketry since physicist Robert Goddard has used chemical fuel. The physics was derived by Russian visionary Konstantin Tsiolkovsky in 1903. The equation that bears his name conveys the brutal truth of how much fuel is required to accelerate a space payload. With Apollo, 95 percent of the mass of the giant Saturn V rocket was burned to get the remaining 5 percent into orbit. The sleek modern rockets of SpaceX and Blue Origin are reusable, but they're still cousins of the Saturn V. Suppose you wanted to send a toothpick to Alpha Centauri. Astronomer Caleb Scharf has done the math.[8] To get it there in a century at 4.4 percent of the speed of light would take an amount of chemical fuel that exceeds the mass of the observable universe by a large factor! We're nowhere near being able to sufficiently miniaturize a fission reactor, and the progress on fusion reactors has been glacial. Other technologies like ramjets are still unproven, and the most efficient fuel of all—antimatter—is a pipe dream.

The best answer is a light sail. Photons carry momentum as well as energy, and when they hit a surface, they give it a small push (twice as much if they're reflected instead of absorbed). Solar sails are powered by the Sun, and this technology was first successfully tested by the Japanese space agency in 2010 with a square sail 14 meters (46 feet) across. NASA and the Planetary Society have also had successful test flights of smaller solar sails. Solar sails deliver a modest but steady thrust and they're likely to be used in the next few decades to propel small probes around the solar system.

In the case of Breakthrough Starshot, the nanobots will first be put into Earth orbit, and once there, will deploy solar sails. An array of lasers on the ground will beam 100 billion watts of energy to accelerate them to 20 percent of the speed of light.[9]

A cascading series of engineering hurdles make the task extremely challenging.[10] The speed to which the probes must be accelerated, one-fifth the speed of light, is a prodigious 215 million kilometers per hour (130 million mph). That takes a phenomenal amount of energy: 100 billion watts is equivalent to a billion 100-watt light bulbs. Lasers exist with sufficient power, but they must beam their energy from the ground to probes in high Earth orbit. Precise focusing and phasing of the laser light is required. The "kick" to the probes must be delivered quickly, within ten minutes, or the energy being beamed to the sails will diffuse out and lose its effectiveness. This much acceleration, which is equivalent to 60,000 g—here, "g" is the acceleration due to gravity at Earth's surface—will create enormous stress on the sails. It will take the probes to 2 million kilometers from Earth, at which distance the sail will appear the same size as a DVD on the Moon. Delivering all the energy creates another problem. So they aren't too heavy, each 4-meter (13-foot) wide sail must be no more than a hundred atoms thick, but they also must be highly reflective. If they absorb more than a tiny fraction of the laser light, they will be heated up and destroyed.

StarChips is the project's name for the probes that will be sent to Proxima Centauri. A StarChip is more like a nanobot than it is a spacecraft, at less than a centimeter across and weighing around a gram. Crammed into this minute package are a camera, electronics to transmit and receive signals, photon thrusters, and a power supply, likely based on a slowly decaying radioactive source such as plutonium-238. The project engineers estimate that if StarChips are mass-produced, the price will come down to that of a smartphone. A thousand StarChips will be needed since there will be attrition due to hits from interstellar dust particles on the long journey. They'll be launched into Earth orbit aboard a "mothership."

Getting nanobots to a nearby star is the central issue, but what about getting the data back? The best idea now is to embed a laser transmitter into each sail. Kevin Parkin is the architect of this technology. He won the Korolev Medal from the Russian Federation of Astronautics for his innovation in thermal propulsion using microwaves. Most spacecraft communication uses microwaves, but as Parkin notes, "Relative to microwaves, lasers have a thousand-fold shorter wavelength, and so form a much tighter beam from Alpha Centauri to Earth. . . . The advantage of transmitting 100 watts over the full area of the sail is that the Earth-based receiver will shrink to a 30-meter telescope, something that is likely to be around in a decade or two." StarChips will be sent in waves and not all at once, so information could be passed back down the chain, like a bucket brigade at a fire. That means StarChips that have already passed through the Alpha Centauri system can relay information to those still on the way. Parkin also suggests including a distributed algorithm that would allow StarChips to function in tandem and with some autonomy since the time needed to make any decisions over interstellar distances is incredibly slow. As he says, "The advantages to doing that are enormous—the entire system could be scouted and mapped before the first data ever reaches the Earth. Notionally, the first StarChip might spot a distant planet as a point of light that moves between images, and on that basis, constrain its orbit so that the next StarChip can maneuver to pass at closer range, resolving surface details. Subsequent StarChips can build up maps, track surface features, and discover most of the planets and moons in the system over time."[11]

Another concern is the speed of the flyby. At 20 percent of the speed of light, StarChips will have just minutes to gather their data before sailing out of the far side of the star system. You might have noticed that with Breakthrough Starshot, the science seems to be buried by the technology. Why not just build a big optical telescope in space, study the planets for months, and get much more information than a "smash-and-grab" mission? There's a clever answer here as well. Gravitational assists can be combined with photon pressure to decelerate an incoming probe, and fling it around a star or bring it to rest. That would allow a leisurely study of any planets that were found there. Autonomous control would allow the probes to explore habitable planets around all three stars in the system.[12]

As they work through their long list of engineering challenges, the project engineers are banking on a continuation of Moore's law to let them cram more sensors and electronics into a tiny payload.[13] Analogous exponential progress is being made in the rising power and falling cost of lasers as well as the number of pixels that can fit into a camera of a particular size and weight. Imagers with two million pixels that weigh less than a gram exist now, but by the time of a launch, those numbers could be a hundred million pixels and a tenth of a gram. Extrapolation is not a strategy, however, and it's unnerving that Breakthrough Starshot is depending on so many different projections of current technologies.

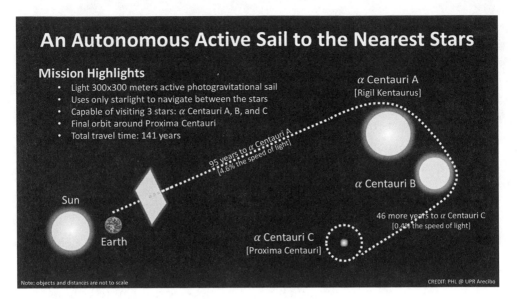

Figure 16.1
Breakthrough Starshot intends to use small robotic probes, propelled by lasers on Earth and solar sails, to reach the nearest star system. With an autonomous design, the mission could potentially reach habitable planets around any of the three nearest stars. The current goal is technology to reach 20 percent of the speed of light, reducing the travel time from ninety-five to twenty years. (Credit: Planetary Habitability Lab, University of Puerto Rico, "Sailing to the Nearest Stars," 2017, https://phl.upr.edu/press-releases/sailing-to-the-nearest-stars.)

With technology improving so rapidly, there's even an argument to procrastinate, as slow spacecraft might be passed by another mission sent with more advanced propulsion. Calculation of this trade-off shows that for most rates of change of progress, waiting is not justified.[14] Avi Loeb, chair of the Advisory Committee for the project, admits the challenges, but calls for action now: "You cannot start the journey without the first steps. Even the critics who argue that the necessary technologies don't exist yet will admit that you have to start somewhere; you have to have a dream. Otherwise, we'll never leave the Earth, and we will be left to our fate."[15]

Milner jump-started the project with $100 million in 2016. The eventual price tag is claimed to be $10 billion, but it might be much higher. The project's management style awkwardly blends NASA's slow and hierarchical approach with the Silicon Valley ethos of putting a bunch of smart people in a room, giving them a long-term goal, and getting out of the way. One committee member, James Benford, says the charge is to "give us next week and five years from now, and we'll figure out how to connect the two."[16]

With a launch date of 2036, the probes will reach their target in 2056, and we'll begin receiving data four years later. The cameras are expected to have the resolution to see not just indications of biology but also large-scale artificial structures if they exist. Mark your long-term calendar for 2060 and a reality show unlike any other.

17
E.T. Phone Home

With billions of habitable worlds in our galaxy, only a pessimist or a killjoy can suppose that all of them are dead. The most prevalent form of life on our planet, and probably on exoplanets, is microbial. Finding microbes beyond Earth would be a dramatic scientific discovery and front-page news, but it probably wouldn't raise the pulse of the public for long.

Four billion years after life on Earth started, evolution has created one species that's intelligent enough to create sophisticated technologies, begin to understand the wider universe, and travel into space. It's natural to ask, Has it happened anywhere else?

The search for extraterrestrial intelligence (SETI) is an ongoing worldwide project that's looking for artificial signals that could only be produced by a sentient creature. SETI makes a series of assumptions. It vaults over all the uncertainty in the path from microbial life to intelligence and technology, and assumes that if it happened here, we're unlikely to be the only creatures with these capabilities. Yet it's a logical fallacy to infer anything general from the evolutionary path that led to us. A skeptic might note that it took billions of years for evolution to arrive at humans, and apart from us there is only a handful of truly intelligent species—other primates, dolphins, orcas, elephants, and perhaps some species of birds—among the hundreds of millions that have lived on the planet.

Biologists have long debated the relative importance of contingency and convergence in evolution. Contingency says there are so many random effects and branching points that there's no

guarantee that "replaying the tape of life" would yield anything like
the same outcome. Contingency was most eloquently advocated by
the late paleontologist Stephen Jay Gould.[1] Convergence says that
evolution often takes multiple paths to common outcomes. There
are many examples of convergent evolution, and they include
brains and eyes, which have both evolved separately several times,
in independent evolutionary lines. A nuanced reading of natural
history suggests that contingency and convergence are both important.[2] The broad envelope of evolution might be similar on another
planet, but the probability of a species like *Homo sapiens* is remote.[3]
SETI researchers fall back on statistics. They contend that the vast
number of potential biological experiments on myriad earthlike
planets increases the odds that there's someone to talk to.

Some*one* or some*thing*. Two other SETI premises relate to technology and communication. Those who are searching admit that
the acronym is misleading; it's more appropriately referred to as
the search for extraterrestrial technology. Scientists don't agree on
how to define intelligence here, let alone beyond Earth.[4] Humans
developed technology in stages: stone tools two million years ago,
hunting weapons fifty thousand years ago, farming six thousand
years ago, the wheel four thousand years ago, steam engines three
hundred years ago, radios and rockets a hundred years ago, and
nuclear power and computers seventy years ago.[5] We see in this arc
an acceleration, and the radio transmitters and lasers preferred for
SETI have only existed for a tiny fraction of the span of the human
species. And it's possible to have intelligence without technology.
Dolphins and elephants might be highly intelligent, but they'll
never point telescopes at the stars or build computers.

Communication is an even trickier proposition. SETI is looking
for signals that aren't of natural origin, but that supposes we can
recognize all the signals that might be generated by astrophysics in
a universe we still don't fully understand. We can't communicate
with chimpanzees, with whom we share nearly 99 percent of our

DNA. How likely is it that we could communicate with an alien species of unknown function and form?

Despite critiques and uncertainty, scientists like to do science. The modern era of SETI was launched with a 1959 paper by physicists Giuseppe Cocconi and Philip Morrison. Its last sentence makes the case: "The probability of detection is difficult to estimate—but if we never search, the chance of success is zero."[6]

The history of SETI shows how it has been framed by the technology of the time. In the first half of the nineteenth century, it was thought that there might be life on Mars or Venus, and the communication was intended for those planets. German mathematician Carl Friedrich Gauss proposed cutting down Siberian forests in the shape of a giant triangle, and astronomer Joseph von Littrow suggested filling a thirty-kilometer (nineteen-mile) diameter circular canal with burning kerosene. Neither of these ideas was ever carried out. Later in the century, when electric lamps were becoming widely used, French inventor Charles Cros imagined using a parabolic reflector to focus light on Mars and Venus.[7] Percival Lowell thought he saw canals on Mars, and his imagination was probably fueled by the boom in canal building during the nineteenth century, culminating in the great constructions at Suez and Panama. Early in the twentieth century, two radio pioneers were interested in using their new technology for interplanetary communication. Nicola Tesla and Guglielmo Marconi each claimed to have detected radio emissions from outer space.[8] By midcentury, advances in radio technology led to the transmitter power of tens of megawatts, putting the stars within reach. The modern era of SETI began.

Frank Drake was fresh out of graduate school at Harvard when he took a job at a new national radio observatory being built in West Virginia. He'd moved from the cosmopolitan confines of Cambridge, Massachusetts, to the rolling hills of Pocahontas County, one of the most isolated regions of the United States, but he was happy because the director was letting him use a large radio telescope to fulfill a childhood dream. "One of the first things that came to my mind was, could this telescope be used to detect other civilizations? So, I did the calculations as to how far away a civilization could be that was sending out radio signals at the same level we were at that time, and we could detect the signals. It turned out to be about 10 light-years. Well, that got interesting. If it had been a tenth of a light-year, we would have had to say forget it, there are no stars within that distance. But within 10 light-years, there are almost a dozen stars. It was not a crazy idea to search for radio signals from those nearby stars, because if any of them were just like us—it didn't have to be a super civilization, just another civilization using no better technology than we had—then we could find them."[9]

A year after Cocconi and Morrison's seminal paper, Drake pointed an eighty-five-foot radio telescope at the bright stars Tau Ceti and Epsilon Eridani. He called the effort Project Ozma, after the Queen of Oz in L. Frank Baum's celebrated books for children. Through a 150 hours of listening, all he heard was a steady crackle of radio noise.

Project Ozma displayed some of the choices researchers face in their quest. SETI is another "needle in a haystack" situation. There are billions of stars in the galaxy, billions of radio frequencies to search, and infinite time that could be devoted to the experiment.

Drake chose to observe two nearby stars, but he could have split up the time and observed a larger number of more distant stars. Therefore sensitivity trades off against the number of targets. In a radio receiver, the signal processing scales with the number of channels. Therefore a large total frequency window trades off against the frequency window for a single channel. Drake reasoned that he should look for signals that were narrow in frequency, with a pattern designed to convey information, since only an artificial signal can be as narrow as 1 hertz. Then there's the huge amount of frequency "real estate," ranging over a factor of 100,000, from marine radio at 100 megahertz, to AM radio at 1 megahertz, FM radio at 100 megahertz, cell phones at 1 gigahertz, and radar at 10 gigahertz. Most radio SETI experiments have been done in a quiet part of the electromagnetic spectrum near 1.5 gigahertz. It's a clean window through the universe; at lower frequencies there's noise from the Milky Way, and at higher frequencies there's noise from the big bang. Moreover, there are fundamental transitions of hydrogen and the OH– ion in this region. A spectral transition is a difference in energy between two energy levels in an atom or ion, so an atom or ion absorbs and emits electromagnetic radiation at frequencies that match the transition energies. Barney Oliver recognized hydrogen and OH– as the ingredients of water, the elixir of life. He dubbed this region of the electromagnetic spectrum the "water hole." Oliver was the vice president of engineering for the Hewlett-Packard Company before becoming chief of NASA's SETI program. "Where shall we meet our neighbors?" he asked. "At the water-hole, where species have always gathered."[10]

SETI has had a rocky road thanks to certain members of the US Congress. In 1981, Senator William Proxmire gave it the Golden Fleece Award as research that "wastes taxpayer dollars," and its funding was terminated. But Sagan persuaded Proxmire of its value and he relented. Then in 1993, Senator Richard Bryan ridiculed SETI again and single-handedly changed NASA's charter to prohibit it looking for "little green men."[11] This was rich since Bryan represented Nevada, and he'd shamelessly lobbied for congressional funds to upgrade the interstate highway near Area 51 and rename it the "extraterrestrial highway."

After that, SETI researchers continued mostly without government funding. The SETI Institute was founded as a nonprofit corporation in 1984. Its original staff of two has grown to over a hundred scientists and dozens of staff members. As radio technology progressed, the searches improved dramatically. The bandwidth of spectrum analyzers grew from a hundred thousand channels to millions and then billions. Radio telescopes around the world participated. The Allen Telescope Array in northern California was instigated with a gift from Microsoft cofounder Paul Allen. Originally planned

Figure 17.1

Some of the radio dishes of the Allen Telescope Array in Northern California. A total of forty-two dishes are distributed over a kilometer, each able to detect narrow band radio signals in the frequency range one to fifteen gigahertz. This array has sped up radio SETI by a factor of a hundred. (Credit: S. Shostak/SETI Institute, 2018, Allen Telescope Array Gallery, https://www.seti.org/allen-telescope-array-gallery#&gid=1&pid=4.)

to include 350 six-meter (twenty-foot) dishes, it had to "hibernate" due to a funding shortfall in 2011, and it's currently stuck at 42 out of 350 dishes.

In 2015, SETI got a huge boost with $100 million from Russian billionaire Yuri Milner, who we encountered as the philanthropist behind the quixotic Breakthrough Starshot project. He's also proposing a mission to try and detect microbes in the icy geysers of Saturn's tiny moon Enceladus. Breakthrough Listen is delivering thousands of hours on two of the largest radio telescopes in the world, turning the current stream of radio data into a torrent.[12] The project reported its first results in 2020. (Spoiler alert: no aliens were found.) No artificial radio signals were detected from twenty stars within five hundred light-years. All were in locations where they would see Earth transiting the Sun. The search was sensitive to radar transmitters like the one at Arecibo Observatory. Hence we know that such transmitters are not being widely used by aliens anywhere near the Sun.[13]

If SETI is a search for a needle in a haystack, the efforts so far have only inspected a few handfuls of straw. These scientists need much greater sensitivity and sky coverage. Their ideal radio telescope would have all-sky coverage and parse billions of frequency channels at once.[14] That telescope is being constructed. It's called the Square Kilometer Array, and will be in Australia and South Africa, with a price tag of $3 billion. The Square Kilometer Array will have fifty times the sensitivity of any existing radio facility. It was designed for general-purpose astronomy, not just SETI, but it will turbocharge the search for radio signals from advanced civilizations.[15]

Most SETI searches are done at radio wavelengths, but some researchers speculated that alien civilizations might use powerful lasers to communicate at optical wavelengths. The problem is that the signal would have to outshine the light of the planet's star. But our best lasers are now so powerful they can exceed the power of the Sun in a short interval like a nanosecond. The method is to observe the light from nearby stars and sample it rapidly so that a pulsed signal would stand out above the steady light from the star. Several optical SETI experiments are in progress.

Radio and optical surveys so far have been a mixture of targeted searches of nearby stars with high sensitivity and wide-area searches of many stars with low sensitivity. If ET transmitters are common and relatively weak, a star-by-star survey will work best. If alien rare beacons are rare but powerful, a wide-area survey will succeed first. Computer simulations of both scenarios suggest that the first signals we detect will come from rare, powerful transmitters at large distances across the galaxy.[16] With such an enormous effort, what's the result?

SETI has not succeeded. The lack of signals from ET over sixty years has been called the "Great Silence." Why haven't SETI scientists just given up?

Drake has an answer: "Use a well-known equation and put in the parameters as we know them. A reasonable lifetime of civilizations is 10,000 years, which is actually much more than we can justify with our own experience. It works out one in every ten million stars will have a detectable signal. That's the actual number. That means, to have a good chance to succeed, you have to look at a million stars at least—and not for ten minutes—for at least days because the signal may vary in intensity. We haven't come close to doing that. We just haven't searched enough."[17]

Jill Tarter is also not the kind of person who quits. She was the first scientist to work at the SETI Institute and became the director of its Center for SETI Research. As a graduate student, she was inspired by NASA's Cyclops Report, which laid out an ambitious plan for detecting intelligent life beyond Earth.[18] She recall that "it was a dense report, but I got hooked. I read it from cover to cover. I was so excited that I happened to live in the first generation of human beings who could try and answer this question. For millennia, all we could do was ask the priests and philosophers; the answers we received were all in tune with someone's belief system. This was an opportunity for scientists and engineers to study the question experimentally, and I thought that was fantastic."[19]

Tarter has won many awards in her career, and has been a vocal and inspirational role model for women in science. The heroine Ellie Arroway in Sagan's book *Contact* was based on her.[20] Tarter recalled working with Jodie Foster to put the character on film: "That was a wonderful experience. We had phone conversations before the movie started shooting and then I went down and worked with her at Arecibo. She's a brilliant, wonderful, amazing woman. It was a great deal of fun, and we formed a friendship."[21]

The scientists searching aren't deterred because their detection capabilities are improving rapidly. It's difficult to know what's good enough to detect what might be out there. Seth Shostak, senior astronomer at the SETI Institute, likes to use an analogy from history: "Christopher Columbus's wooden ships were slow, uncomfortable and dangerous. If he had waited 500 years, he'd have crossed the Atlantic in a couple of hours, eating bad food. Wooden ships were good enough in that case. You don't know where the threshold is."[22] Within twenty years, SETI scientists will have done surveys sufficient to detect analogs of our radio transmitters and lasers operating on planets around the nearest million stars. This is an improvement by a factor of a thousand on what we can do now, and a factor of a billion on Drake's first experiment in 1959. If the silence continues, it could mean that advanced civilizations are much rarer than we think, just not interested in communication, or don't exist. If we do detect a signal, as envisaged in the book and movie *Contact*, it will be an Earth-shaking event.

Ever the optimists, SETI researchers have a set of protocols for what happens if we do get a ping from the cosmos. The essentials are to verify that it's truly an extraterrestrial source, then announce its discovery to the world, and don't reply without first seeking international consultation and approval.[23]

Meanwhile, many people think scientists are late to the party because we've *already* been visited by aliens.

Unidentified flying objects (UFOs) are a worldwide phenomenon where people report mysterious moving objects in the sky that don't seem to have any conventional explanation. Ground zero for modern UFO sightings was the incident in Roswell, New Mexico, in July 1947. There's now an entire "industry" revolving around UFOs, alien abductions, and the associated government cover-ups of the information about aliens visiting Earth.[24] Almost all scientists are skeptics because of the lack of verifiable, physical evidence and the fact that so many UFO sightings have conventional astronomical explanations.[25] In 2021, the subject was in the news after the release of US Navy videos showing aircraft moving in strange and seemingly unphysical ways. A report on 144 of the sightings by the Office of the Director of National Intelligence was unable to classify most of them, but it declined to say that they were visits from aliens.[26]

I was drawn into the "rabbit hole" of UFO sightings after writing an article for the online magazine *The Conversation*, where I expressed skepticism that there was any credible evidence of visits from aliens. In a dozen other articles for that magazine, I'd rarely attracted more than a hundred thousand readers and a dozen comments. But this article drew a torrent of reaction: half a million readers and over five hundred comments. Many were critical, and a few were abusive; I'd rattled the hornet's nest of the "true believers." They regaled me with UFO sightings along with stories of contact and abduction, and the experience left me numb and exhausted. It's no wonder that most professional astronomers are reluctant to engage in the subject.[27]

Some fraction of UFO sightings needs further investigation, but both UFO reports and any claimed SETI detection should be subject to the level of judgment made popular by Sagan: extraordinary claims require extraordinary evidence.

18
Energy Footprints

The search for extraterrestrial intelligence, SETI, is inherently anthropocentric. It assumes intelligent aliens are interested in communication and use technologies familiar to us. SETI thus far is also severely limited. In terms of the radio searches, I return to the analogy of trying to find fish in a body of water. If the proportion of the universe SETI has searched so far is represented by a glass of water, the rest of space is equivalent to the water in all of Earth's oceans.[1]

An alternate strategy is to look for signs of a civilization's energy use or artifacts it might create. Comparing this with the astrobiologists' approach of looking for biosignature gases in exoplanets' atmospheres, Jill Tarter dubbed the signs of an advanced civilization visible from afar as "technosignatures."

Tarter says, "If we can find technosignatures—evidence of some technology that modifies its environment in detectable ways—then we will be permitted to infer the existence, at least at some time, of intelligent technologists. As with biosignatures, it is not possible to enumerate all the potential technosignatures of technology-as-we-don't-yet-know-it, but we can define systematic search strategies for equivalents of some 21st-century terrestrial technologies."[2]

Just like biosignatures, technosignatures would be a detectable sign of existing or extinct life. They represent any sign of technology that we can use to infer the existence of intelligent life, including the traditional communicative SETI tools of narrow band radio signals and pulsed lasers. The process is to hypothesize a technosignature, motivated by life on Earth, and then design a search that considers its detectability and uniqueness. Many proposed techno-signatures involve the manipulation of energy from bright natural sources. Technological life might spread through the galaxy, so its tracers may be found far in time and space from where that life originated. As the 2018 report from a NASA workshop framed it, "Compared to biosignatures, technosignatures might be more ubiquitous, more obvious, more unambiguous, and detectable at much greater (even extragalactic) distances."[3]

Earth has undergone violent changes in the distant past. Major episodes include intense volcanism and heavy bombardment by meteors soon after the planet formed, two "Snowball Earth" events 2.3 billion and 650 million years ago when the entire surface might have frozen, intense volcanism 250 million years ago, a catastrophic impact 65 million years ago, and a period of dramatic warming 55 million years ago. Lovelock's Gaia theory suggests that biology is intimately linked to the physical environment. When the environment changed suddenly, 250 and 65 million years ago, life suffered mass extinctions and the genetic landscape was redrawn. When cyanobacteria started building up oxygen in the atmosphere, the result was fatal to many existing species, but it yielded the biomarker that astronomers will use to look for microbial life on other planets.

When the first modern humans emerged about 200,000 years ago and radiated out of Africa 50,000 years ago, they adapted to Earth's changing climate. Over that time, the global temperature varied by 10°C (18°F) through a series of changes called glacials and interglacials.[4] With a population of just a few million, we humans had a light footprint on the planet. Since the end of the last Ice Age 12,000 years ago, a time that geologists call the Holocene, we've lived in a "Garden of Eden" with a remarkably stable climate. Humans flourished, they invented agriculture, and the population began to grow. But recently we've seen a crescendo of human activity that's profoundly affecting Earth's geology, climate, and ecosystems. Many geologists call this epoch the Anthropocene—although this term isn't officially recognized.[5] Changes started to be non-linear with the Industrial Revolution, but the rate of change has increased in the past 60 years.[6] The "Great Acceleration" is marked by a growing population, industrial production, agricultural chemicals, and detritus from the nuclear age.[7] We're witnessing the most profound transformation of our relationship with the natural world in the history of humankind.

As pointed out previously, although we wrestle with climate change and environmental destruction, the human impact on the planet wouldn't be apparent to a remote observer. Intelligent extraterrestrial beings far away wouldn't be able to detect our large cities, great construction projects, or even injection of carbon dioxide into the atmosphere. We've been leaking radio and television signals into space for seventy years. That means a spherical wave of those transmissions has washed over several thousand stars and their habitable planets. Have the aliens heard the first episodes of the TV show *I Love Lucy*, and decided we're weak and frivolous creatures who deserve to be conquered? It's moot because I worked out that these signals faded into the noisy radio hum of the cosmic background radiation before they ever left the our solar system, and I had to break the news to NPR listeners.[8] We're not that impressive.

This awareness of our modest impact leads to radical speculation. Maybe we're not the first industrial civilization on our planet.

Astrophysicist Adam Frank and climatologist Gavin Schmidt call it the Silurian hypothesis.[9] Earth is a restless planet, with erosion and tectonics leaving little ancient surface visible today. Our great cities will be subsumed in millions of years, and fossilization preserves a tiny fraction of dead animals. Millions of years from now, the geological record will be mute to our existence. Only a civilization that was long-lived and intensive in its use of energy would be detectable. Silurians were an ancient race of technologically advanced, reptilian humanoids from a 1970 episode of the classic BBC TV series *Doctor Who*. The Silurian hypothesis is relevant to astrobiology. We can imagine how hard it would be to detect a transient, technological civilization on a distant exoplanet if we couldn't detect a prior one on our own planet!

Looking at the Anthropocene epoch through the lens of astrobiology can help us make sense of the future evolution of life on Earth and potential evolution of technological life elsewhere in the universe. Nikolai Kardashev was the visionary who was the first to speculate about the capabilities of civilizations that had left us in the dust. Kardashev had a challenging early life in Stalin's Russia. At age five, his father was executed, and his mother was imprisoned, so he was raised by his aunt. At age twelve, he began hanging out with young astronomers at the Moscow Planetarium. For many years, his mentor was Iosef Shklovsky, whose 1962 book *Universe, Life, Intelligence* inspired the young Sagan and marked the start of astrobiology in the Soviet Union. Kardashev made major contributions to radio astronomy and predicted pulsars, but he is best known for the Kardashev scale of alien civilizations. Given the brutality he saw growing up, Kardashev was surprisingly optimistic that life in the universe had an ethical basis: "The concepts of morality and goodness are universal, like the Pythagorean theorem. Civilizations do not survive if they do not follow these concepts."[10]

His scale measures how technologically advanced a civilization is by how much energy it can use. As proposed in 1964, it had three categories.[11] A Type 1 civilization uses all the energy available on its planet, which for the solar radiation that falls on Earth is 2×10^{17} watts. A Type II civilization harnesses all the radiation emitted by its star. Energy use is the luminosity of the star; for the Sun that is 4×10^{26} watts. Comparing these two numbers, we see that Earth intercepts half a billionth of the Sun's light. A Type III civilization is unimaginably advanced, harnessing energy at the scale of a galaxy. For the Milky Way, this energy usage would be 4×10^{37} watts. Kardashev thought this was the limit, but futurists like Michio Kaku have thrown caution to the wind and imagined Type IV civilizations that could command the power of a billion trillion suns—cultures spanning the breadth of the universe. The ultimate concept is a Type V civilization, transcending its universe of origin to be the master of the multiverse. Ignoring for a moment the grandiosity of the last two categories, they describe entities with capabilities so enormous that they're indistinguishable from deities.

Where are we in this hierarchy? We're nowhere near gods; instead, we're down in the foothills below Type I. World power consumption is 20 terawatts or 2×10^{13} watts. We use energy at one-ten-thousandth the rate it falls on our heads from the Sun.

We might be minnows in the sea of technological civilizations, but we're at a pivotal stage of our growth. The Kardashev scale makes the crude assumption that technology improves as energy use grows, and more advanced civilizations will naturally need and use more energy. It's a brutalist way of thinking, where progress occurs by mastering a star, then a galaxy, and finally the universe. It's as if we measured someone's progress by the amount of money they earned. Money is an easy way to keep score, but it ignores the value of that person's ideas, cooperation, and ethics. As we navigate the Anthropocene, we can't simply grow our way out of the challenges we face. Gaia tells us we can't impose our will on the biosphere. Even without growth, it's unclear if any civilization as energy intensive as ours can survive for centuries.

A group led by Frank at the University of Rochester has considered sustainability on a planetary scale. The group uses a thermodynamics perspective, looking at fundamental limits on how energy can be generated as well as the consequences on the biosphere of using that energy.

"Our premise is that Earth's entry into the Anthropocene represents what might, from an astrobiological perspective, be a predictable planetary transition," the group members write. "We explore this problem from the perspective of our own solar system and exoplanet studies."[12]

The group developed a classification scheme for evolutionary stages of worlds based on the thermodynamics being out of
equilibrium—the planet's energy flow being out of synch, as the presence of life can cause. The first class represents worlds where the
energy flow is in equilibrium, and examples include airless worlds like
the planet Mercury and Earth's Moon. The second class of planets has
a thin atmosphere with greenhouse gases but no life, like Mars and
Venus in their current state. The third class of planets has a thin biosphere and some biological activity but not enough to alter the evolution of the whole planet. Early Earth may have been such a world
when photosynthesis and other wide-scale biological activity began
to strongly affect the energy flow of the planet. Perhaps this was once
also true of Mars. The fourth class has a thick biosphere, with extensive cycling of material between the crust and the atmosphere. Now
Earth is transitioning to the fifth class, where it's profoundly affected
by the activity of an advanced, energy-intensive species. Frank and
his colleagues offer the sober warning that such a state is only stable
and sustainable "if humanity successfully manages a transition to
an energy system based entirely on solar energy."[13]

Civilization on Earth inefficiently uses a small fraction of the
Sun's energy and releases it as waste heat energy into space. This
thermodynamic principle will hold anywhere in the universe. An
advanced civilization using "free" energy from its star will reveal
itself by the excess heat it generates.

The hypothetical tool for utilizing all the energy output of a star
is called a Dyson sphere, a physical structure built around the
star that captures the energy and beams it to the civilization on its
planet. This idea came from the fertile mind of Freeman Dyson, a
British-born math prodigy who made outstanding contributions to
theoretical physics.[14] The sphere is often described as a solid shell
surrounding a star, but Dyson knew such a design was mechanically
impossible. It's better understood as a "swarm" of independent solar-
powered satellites orbiting in a dense formation around the star. As a

young graduate student, Dyson wrote a paper on quantum electro-dynamics that his colleagues thought was worthy of a Nobel prize. He didn't bother to finish his PhD, but within two years he was a professor of physics at Cornell. Realizing that he didn't enjoy teaching, he spent most of his career at the Institute for Advanced Study in Princeton, New Jersey. He wrote research papers and books well into his nineties, and had an iconoclast's mindset, arguing against the scientific consensus on climate change. His friend and physics Nobel Prize winner Steven Weinberg reflected, "I have the sense that when consensus is forming like ice hardening on a lake, Dyson will do his best to chip at the ice."[15]

Even with a Dyson sphere capturing a star's output, there would inevitably be energy that escapes into space, or waste heat, in the

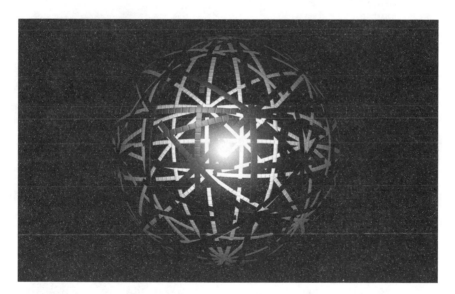

Figure 18.1
The archetypal technosignature is a Dyson sphere, named after the eminent physicist Freeman Dyson. This artist's impression shows a loosely connected set of solar energy collectors arranged around a star to collect most of its energy. (Credit: Sentient Developments, in "Technosignatures and the Search for Extraterrestrial Intelligence" by M. Kaufman, 2018, NASA Astrobiology, https://astrobiology.nasa.gov /news/technosignatures-and-the-search-for-extraterrestrial-intelligence/.)

form of infrared radiation. In practice, it is hard to distinguish a civilization's waste energy from extra infrared energy due to various natural processes. For example, dust around a star reradiates some of the star's energy in the infrared. A search of twenty-five thousand stars for excess infrared emission failed to turn up any convincing signs of Dyson spheres.[16] So far, there's been no convincing claim of the energy footprint of a civilization. There was excitement several years ago over the dramatic and irregular light variations of Tabby's star, which was named after its discoverer. Some speculated that the variations might be due to an orbiting megastructure built by an alien civilization, like a Dyson swarm. SETI scientists turned the Allen Telescope Array toward Tabby's star in 2015, but they saw no artificial radio signals, and there are more conventional explanations for the variations.[17]

In a field that is limited only by imagination, many technosignatures have been discussed, along with the ways they might be detected. Conventional examples include detecting the light from a worldwide city or industrial pollutants like nitrogen oxide in a planet's atmosphere. Extravagant illustrations include stellar thrusters— where a civilization uses stellar energy to accelerate a star and everything orbiting it in any direction—and civilizations that use antimatter as a power source.[18] These would potentially be easy to find and interpret. There's even been speculation about civilizations that can make stars and even galaxies disappear, bringing to mind science fiction writer Arthur C. Clarke's adage, "Any sufficiently advanced technology is indistinguishable from magic."[19] A new generation of large telescopes and observatories in space has the means to detect several advanced technologies for a hundred million stellar targets.[20]

As an instance of how the debate is framed by our current limitations, consider gravity waves. Modulating space-time ripples would be an excellent way to send information over galactic or intergalactic distances. Their signal strength diminishes more slowly with distance than light or radio waves. They're not scattered or absorbed by gas or stars, so they travel unimpeded. They have no background noise to compete with. The only disadvantage is that they require an enormous amount of energy to create a detectable signal. But that last statement is only true of our current technology. A more advanced civilization may have no trouble generating them, and they may have long ago discarded lasers and radio transmitters as childish toys. We've only been able to detect these waves in the past few years. We would have missed gravitational wave signals that arrived thousands of years ago—or even a couple decades ago.

On a more local scale, if we're sending probes into space, others may be doing so too. What if there are alien artifacts in our solar system, right under our noses?

To be easily detectable, the probes would have to be big and obtrusive. Think of the monolith in *2001: A Space Odyssey*, an evocative concept of Clarke. If the aliens sent miniature probes to visit us, as we do around the solar system and might eventually do to Proxima Centauri, our odds of detecting them are low. The "sentinel hypothesis" was proposed back in 1960 by engineer Ronald Bracewell (these probes are sometimes called lurkers).[21] Searching for sentinels is yet another needle in a haystack project. Any artificial object with flat, metallic surfaces, however, will stand out since light or other electromagnetic radiation reflected off a flat surface is polarized (with all the waves' vibrations occurring in the same plane). Astronomers have tools to detect polarized radiation. Fifty thousand solar systems passed close to ours in the past 4.5 billion years, any of which could have easily investigated us.

In 2017, astronomers detected a cigar-shaped object that did come from outside our solar system. Called Oumuamua, this first-known interstellar visitor caused quite a stir. Oumuamua whipped through the solar system on a high-speed, hyperbolic orbit that had astronomers scratching their heads. Most now think it's a natural object—an icy piece of a planet like Pluto.[22] We met Avi Loeb when he argued that life could spread through the universe of hyper-velocity stars. This time he caused a splash with the maverick view that Oumuamua was an alien artifact. He inverted the standard that Sagan applied to UFOs: "It is not obvious to me why extraordinary claims require extraordinary evidence," and countered with "extraordinary conservatism keeps us extraordinarily ignorant."[23] Even if aliens don't send detectable artifacts, simulations show that modest-speed starships would allow aliens to traverse the galaxy in a small fraction of its age.[24]

If alien civilizations exist at all, we're unlikely to be the most advanced, so their technological capabilities would overshadow ours. That makes it worth putting effort into clever ways of searching for those technologies. The odds of success may be impossible to predict, but the payoff would be immense.

19

The Drake Equation

"We are alone in the universe, or we are not. Either way, the implication is staggering." This quote captures tension at the core of the human condition.[1] We've seen projections of the enormous numbers of habitable planets in the Milky Way, yet we still don't know if any of them host biology. We're unsure what the essential ingredients for biology are, or if biology elsewhere will use the same genetic code or lead to similar organisms as life on Earth.

We don't know if intelligence is inevitable or a highly contingent outcome of evolution, and the role of technology is completely uncertain. Yet the possibility that other civilizations have emerged in the galaxy, some surpassing us in their capabilities, is an intoxicating idea. What intelligence might have evolved there? Do they have lessons for us on how to survive our troubled adolescence as a species? It would be useful to be able to bundle these disparate considerations into one handy formalism. We can; it's called the Drake equation.

The young postdoc walked up to the blackboard and started writing. Frank Drake was trying to organize his thoughts on life in the universe. He was thirty. Two years earlier, he had conducted the first search for radio signals from an extraterrestrial civilization (see chapter 17).

The National Academy of Sciences had asked Drake to organize a meeting in 1961 at Green Bank Radio Observatory in West Virginia. "I was the entire scientific and local organizing committee, but it wasn't a hard task because I invited every person in the world we knew of who was interested in working in this subject—all

twelve of them. And all twelve of them showed up." The partici-
pants included Giuseppe Cocconi and Philip Morrison, authors
of the seminal paper "Searching for Interstellar Communications";
Barney Oliver, the founder and director of Hewlett-Packard laboratories; Dana Atchley, a radio ham and electronics entrepreneur;
Carl Sagan, a young rising star of astronomy; the chemist Melvin
Calvin, who learned that he'd won a Nobel prize during the meeting; and neuroscientist John Lilly, who'd done work on communication with dolphins. In honor of Lilly, the group called itself the
Order of the Dolphin. Lilly was a part of the counterculture movement of the 1960s, in league with Ram Dass, Werner Erhard, and
Timothy Leary. He was a controversial scientist, whose research on
consciousness involved the use of psychedelic drugs and sensory
deprivation tanks.[2]

Drake recalls the genesis of his iconic equation: "As I planned
the meeting, I realized a few days ahead of time that we needed an
agenda. And so I wrote down all the things you needed to know to
predict how hard it's going to be to detect extraterrestrial life. And
looking at them it became pretty evident that if you multiplied all
these together, you got a number, N, which is the number of detectable civilizations in our galaxy. This, of course, was aimed at the radio
search, and not to search for primordial or primitive life forms."[3]
What Drake wrote on the board was the equation that would bear his
name and be a touchstone for astrobiology for the next sixty years.
As he described it, "I didn't have a real value for most of the factors. But I did have a compelling equation summarizing the top-
ics to be discussed. The number (N) of detectable civilizations in
space equals the rate (R) of star formation, times the fraction (f_p) of
stars that form planets, times the number (n_e) of its planets that are
hospitable to life, times the fraction (f_l) of those planets where life
actually emerges, times the fraction (f_i) of those planets where
life evolves into intelligent beings, times the fraction (f_c) of those
planets with intelligent creatures that are capable of interstellar

communication, times the length of time (L) that the civilization remains detectable."[4]

The Drake equation combines factors to estimate the number of intelligent communicable civilizations at any time in the Milky Way galaxy. The first factors are astronomical, and have been measured or estimated: the rate that new stars are forming in the galaxy, and how many habitable planets there are per star. The remaining factors are unknown: Given a habitable planet, how often does life form, how often does it evolve intelligence, and how often does it develop the technology to travel and communicate through space? The final factor is a complete wild card: the longevity of the civilization.

Drake continues: "At the conference, we plugged in our best estimates for each of the factors, and found that the product of the first six factors on the right-hand side came out to roughly the value of 1. Thus, the value of N seemed to hinge solely on the value of L—how long intelligent, communicative civilizations can survive."[5] If the estimates at this first meeting were correct, the Drake equation simplifies to $N = L$, the number of galactic pen pals equals their average lifetime.[6] We can look in a mirror and project the answer. If we destroy ourselves a century after developing bombs and computers, the galaxy is a lonely place. But if we can survive and endure, there will be companionship out there.

What are the best estimates of these six factors, sixty years after Drake rolled out his equation for the first time? The rate of star formation in the Milky Way (R) is three per year, most of which are low-mass red dwarfs.[7] The microlensing surveys discussed earlier suggest at least one planet per star, so $f_p = 1$.[8] Kepler satellite data has found $n_e = 0.5$, or one habitable planet for every other star system. Exoplanets discovered by Kepler, however, are in a nearby region of the galaxy. A full consideration of the challenges to habitability across the galaxy finds a "galactic habitable zone" that is a slender ring in the disk of the galaxy. Too close to the galactic center, and high star densities will lead to excessive supernova explosions and star interactions that disrupt planetary systems. Too far out in the disk and the heavy element abundance is low, so carbon-based chemistry is less likely. The spherical halo of the galaxy also has low heavy element abundance, so life will be scarce there.[9] From this study, $n_e = 0.1$. So far, so good—although the possibility of life on exomoons, as discussed in chapter 12, muddies the water. Ignoring those uncertainties for now, the product of the first three factors, $R \times f_p \times n_e$, is about 0.15.

Then it gets sketchy. Life started soon after Earth formed, suggesting that life is common if conditions allow, so f_l, the fraction of planets where life emerges, is close to 1. Yet we're biased in looking at the planet we already inhabit, so that's not a valid argument. A better case that $f_l = 1$ could be made if we find current or past life on Mars, Europa, Enceladus, or Titan. The probability of life developing intelligence, f_i, is particularly controversial. Eminent biologist Ernst Mayr argued that f_i is much less than one since only one of the several hundred million species that have lived on Earth has become intelligent.[10] Those who favor f_i closer to 1 point to a history of increasing biological complexity and a progression from senses to central nervous systems to brains.[11] Arguments over the value of the fraction of planets with intelligent creatures capable of interstellar communication, f_c, are impossible to resolve. Other mammals may be intelligent, but the ability to communicate through space is highly specific to humans. This much uncertainty leaves no path forward.

Drake's equation is not a normal math tool where you plug in the numbers and get an answer. He has admitted it's more like a container for ignorance.

The main problem is that in any product of numerical factors, each factor has equal importance. But a chain is only as strong as its weakest link. The first three factors are astronomical, and they've been measured with increasing reliability in the past decade. The last four are unknown, though, and different researchers have made seemingly plausible estimates for each of them that range over several orders of magnitude. With only one form of biology to study, our understanding of the pathway from life's ingredients to biology to intelligence to technology is paltry at best. The longevity of any technological civilization could range widely too.

Let's use that factor to illustrate the uncertainty. At one end, Michael Shermer, founder of the Skeptics Society, came up with an estimate of L=420 years based on the duration of sixty historical civilizations on Earth.[12] Many of the civilizations, however, carried on technologies of their predecessors, so they're not independent estimates of L. At the other extreme, astrobiologist David Grinspoon speculated that some civilizations might evolve sufficiently to overcome all threats to their survival. They would essentially become immortal.[13] If L is an enormous number, potentially billions of years, the galaxy may have been accumulating advanced civilizations since it formed. In practice, L may not have a well-defined average but rather a wide distribution. If that's the case, the probability of contact would be set by the long tail of ancient civilizations, just as the probability of detecting a radio signal would be dominated by the rare civilizations with extremely powerful transmitters.

We've got "three measured numbers" times "I don't know" times "not sure" times "it's just a guess" times "haven't a clue." The answer is: haven't a clue. That's a pessimistic and glib assessment, but modifications to the Drake equation have been proposed.

One study provided a statistical generalization of the Drake equation, using random variables as opposed to numbers for each factor.[14] After many pages of dense math, the results were intriguing. The baseline was a set of plausible numbers that multiplied out to N=3,500 communicable civilizations in the galaxy. A statistical treatment shows that the mean number N=4,600 is higher than the 3,500 calculated by the classical Drake equation working with numbers only. The range is large, from a theoretical minimum of zero (or one, since at least humanity exists!) up to tens of thousands. Based on the geometry of the galaxy, the most likely distance to the nearest extraterrestrial

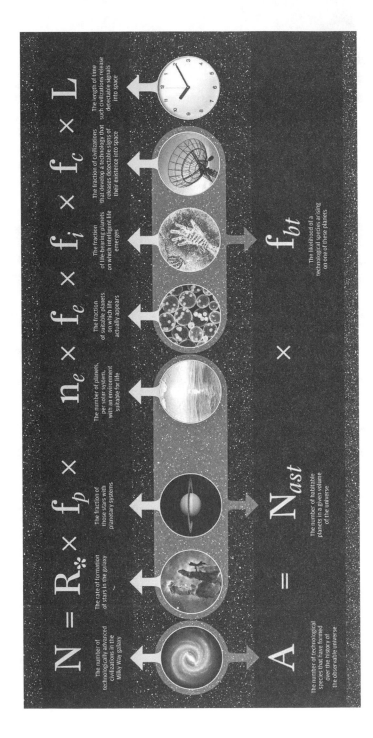

Figure 19.1
The Drake equation combines a series of numerical factors to estimate the number of intelligent, communicable civilizations in the galaxy at a particular time. The first three factors have been measured, but the last four are unknown. (Credit: University of Rochester, in "Are We Alone in the Universe? Revisiting the Drake Equation" by L. Sierra, updated 2021, NASA, https://exoplanets.nasa.gov/news /1350/are-we-alone-in-the-universe-revisiting-the-drake-equation/.)

civilization is 1,900 light-years. Looking at the statistical distribution, the probability of finding extraterrestrials is practically zero up to 500 light-years from Earth, and with 75 percent probability, the distance to the nearest civilization is between 1,350 and 4,000 light-years. That's 16 thousand trillion kilometers (10 thousand trillion miles) away—and a round-trip message time of four millennia. Not exactly neighbors.

Sara Seager has adapted the equation to address the search for bio-signature gases from microbes rather than radio signals from intelligent aliens. She says, "Detectable signatures of gas could mean a lot of things. As human beings, we exhale carbon dioxide. That's our biosignature gas. But that's not useful because carbon dioxide in the atmosphere is naturally occurring. There are other possible gases we could look for. Oxygen is produced by plants and photosynthetic bacteria. We have also considered ammonia as a biosignature gas." For her estimate, she assumes the following strategy: "We'll use the Transiting Exoplanet Survey Satellite to find rocky planets transiting small stars. Then we'll use the James Webb Space Telescope to observe the atmospheres of those planets, during transits or secondary eclipses. The punchline here is that if we're really lucky and everything works in our favor, we will be able to infer signs of life on those planets."[15] Her calculation suggests that a few inhabited planets could reasonably turn up during the next decade.

Other modifications to the Drake equation have been proposed. If civilizations colonize the galaxy, propagating at a certain speed and establishing new sites that survive for a certain time, the single Drake equation becomes a more complex set of three equations. Civilizations might reappear on a planet, letting those planets host multiple sequential civilizations (remember the Silurians!). Moreover, being in a communicative phase and sending dedicated messages are not the same thing. Only a fraction of the civilizations that have the technology to communicate in space might do so.

Astrobiologists sort into "pessimists" and "optimists." Pessimists think intelligent life with our capabilities is a fluke and we're alone. Optimists think there's some inevitability to the progression that led to us, and we have cosmic companionship. Who does the data give succor to, the pessimists or optimists?

Adam Frank and astronomer Woody Sullivan tackled this question. They first noted that the Drake equation applies only to our galaxy. The question "Are we alone?" means considering all 100 billion galaxies in the universe. They reframed the Drake equation into a form that let them ask, Are we the only technological species that has ever arisen? They calculated how unlikely advanced life must be if there has never been another example among the universe's 10 billion trillion stars, or even among our Milky Way galaxy's hundred billion stars. The answer? By applying recent exoplanet data to the universe's 2×10^{22} stars, Frank and Sullivan found that human civilization is likely to be unique in the cosmos only if the odds of a civilization developing on a habitable planet are less than about 1 in 10 billion trillion, or 1 in 10^{22}. "One in 10 billion trillion is incredibly small," said Frank. "To me, this implies that other intelligent, technology-producing species likely have evolved before us. Think of it this way. Before our result you'd be considered a pessimist if you imagined the probability of evolving a civilization on a habitable planet were, say, one in a trillion. But even that guess, one chance in a trillion, implies that what has happened here on Earth with humanity has happened about 10 billion other times over cosmic history!"[16]

Those numbers speak to the optimists, but calculating the odds that other civilizations are around today gives ammunition to the pessimists.[17] The universe is over thirteen billion years old. That means that even if there have been a thousand civilizations in our galaxy, if they live only as long as we've been around—roughly ten thousand years—then all of them are likely to already be extinct. Others won't evolve until we are long gone. For us to have any chance of success in finding another "contemporary" technological civilization, on average they must last much longer than ten thousand years. Given the vast distances between stars and the fixed speed of light, we may never be able to have a real conversation with another civilization.

The take-home message from Frank and Sullivan's work: it is astonishingly likely that we are not the only time and place that an advanced civilization has evolved. As we face our current sustainability crisis, we wonder if other civilizations have reached such a bottleneck and made it to the other side. Scientists can use everything they know about planets and climate to model the interactions of an energy-intensive species with their world, knowing that many situations like this have already occurred somewhere in the cosmos. We can also deploy simulations to determine what leads to long-lived civilizations. Hopefully this will illuminate a path through the darkness we face as a species.

20

The Fermi Question

Enrico Fermi looked at his colleagues and asked, "Where are they?"
They laughed and waited for him to explain where the question
came from.

It was 1950, and Fermi and three physicist coworkers were hav-
ing lunch in the canteen at Los Alamos National Laboratory in New
Mexico. The previous week, they'd talked about a spate of recent
UFO reports in the newspaper, and a cartoon in the same paper
showing aliens stealing trash can lids. Fermi's question seemed to
come out of the blue, but his colleagues guessed he was talking about
extraterrestrial life.[1]

Fermi's rapid-fire intellect was famous among his fellow scientists.
He excelled at both theoretical and experimental physics. He won a
Nobel Prize in 1938 for his research on radioactivity. In 1942, he demonstrated the first controlled nuclear chain reaction and invented a
method to estimate the yield of nuclear bombs for the Manhattan
Project.[2] Physics teachers still rely on his habit of using intuition and
estimation to get a rough answer to a complex problem. He was nicknamed "The Bishop," not because he was religious, but because he
seemed infallible on matters of physics. In this case, his quick mental
calculation involved estimating the incidence of earthlike planets,
probability of life given an Earth, probability of humans given life,
and likely duration of their technology. He did this a decade before
Frank Drake did the first SETI experiment and came up with his equation, and forty-five years before the first exoplanet was detected!

Fermi's question has been called a paradox, but it doesn't fit the formal definition, which usually involves a self-contradictory statement.[3] As we'll see, there are many explanations for the absence of visits from intelligent aliens. Searches for extraterrestrial life or the artifacts of extraterrestrial civilizations are at a primitive stage. SETI employs specific strategies and has surveyed a tiny fraction of the "landscape" for signals. Advanced civilizations might leave a large energy footprint, but they might use energy efficiently and leave no footprint. Their interstellar probes might be small and essentially undetectable. As the aphorism goes, "Absence of evidence is not evidence of absence."

The first and most obvious answer to Fermi's question is that extraterrestrial intelligence is rare or nonexistent. The analysis of Adam Frank and Woody Sullivan that we encountered in the previous chapter achieves its potency by asking how likely is it that we're the only technological species in any galaxy in the universe across all cosmic time. If we're restricted to a contemporaneous civilization, in this galaxy only, the odds are not nearly as impressive.

Evolution may operate with contingency such that intelligence isn't a natural or inevitable outcome.[4] In other words, evolution might be subject to a Great Filter, as discussed in chapter 14, that renders intelligent life much rarer than microbial life. In this case, biology is quenched before it can develop technology. On Earth, the Gaia mechanism has regulated greenhouse gases, maintaining the surface temperature suitable for liquid water and habitability. It took several billion years for complex life to evolve on Earth, and most early planetary environments are unstable—so perhaps in most cases the Gaia mechanism doesn't have time to operate, and life is snuffed out. Four billion years ago, Earth, Venus, and Mars might all have been habitable, but only Earth stabilized with a habitable climate.[5] Another episodic hazard that could act as a filter on any planet is mass extinctions, which interrupted evolution on Earth every few hundred million years. Catastrophes due to meteor

The Great Filter

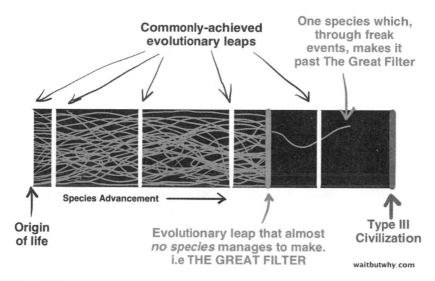

Commonly-achieved evolutionary leaps

One species which, through freak events, makes it past The Great Filter

Species Advancement ⟶

Origin of life

Evolutionary leap that almost *no species* manages to make. i.e THE GREAT FILTER

Type III Civilization

waitbutwhy.com

Figure 20.1

Among the explanations for why we haven't seen any sign of intelligent aliens is the idea of a Great Filter, one or more obstacles or bottlenecks in the way of evolution from simple organisms to intelligence and technology. (Credit: T. Urban, in "The Fermi Paradox," 2014, *Wait But Why*, https://waitbutwhy.com/2014/05/fermi-paradox.html.)

impacts from space, gamma ray bursts, or violent volcanic activity could greatly lower the odds that biology reaches the stage of intelligence and technology.[6] Finally, as on Earth, intelligence need not lead to technology. This may be particularly true on "water worlds," which I explored in chapter 10. On exoplanets that are completely covered with deep oceans, the evolution of land animals who develop tools and technology would be impossible.

If we venture beyond astronomy and biology into sociology, we encounter more arguments as to why we see no sign of intelligent aliens.[7] Here's a smorgasbord.

One possibility is that intelligent life destroys itself. For seventy-five years, the *Bulletin of the Atomic Scientists* has documented our proximity to global catastrophe.[8] Its Doomsday Clock is a countdown to an apocalypse that would spell the end of our species—and it's been close to midnight in the past due to the threat from nuclear weapons. Currently, we're less than two minutes from midnight, with existential threats from nuclear weapons, biological pathogens, and climate change.[9] Even if evolution gives rise to intelligent creatures on another planet, they may attempt to communicate using technologies we cannot yet harness, like neutrinos or gravity waves. Or they may not feel any cultural imperative for colonization or off-world exploration. They may even choose not to interact with us and be just watching instead. This is sometimes called the "zoo hypothesis." More ominously, they may destroy civilizations that arrive at a particular stage of development—in which case our doom is just around the corner. This is called the Berserker hypothesis in a series of science fiction books by Fred Saberhagen.[10]

Here's another way of explaining Fermi's question: alien life may be too "alien" to recognize. It's one possibility that's hard to exclude and conceptualize. There's a strain of anthropocentric thinking that permeates exobiology. Since Earth is the only place we know where an intelligent, space-faring species has evolved, we tend to use that example as indicative of the capabilities of life on other worlds. Life elsewhere might be so disconnected from us in time and space that it's unrecognizable. Imagine a form of biology where the information density is extremely high, so complex organisms could be tiny compared to us. Or imagine life-forms with rapid metabolisms, so the cadence of their life cycle is much shorter than ours.

Another possibility is that biology evolves to the point where intelligence becomes purely computational. Postbiological evolution could give almost unimaginable properties to creatures that leave the limitations of conventional biology behind. This sounds fanciful until we realize that humans are approaching this "membrane" via the development of artificial intelligence, robotics, and replacement of body parts with machinery.[11]

By now, we're firmly in the grip of speculation and should recall writer Mark Twain's wry comment, "There is something fascinating about science. One gets such wholesale returns on conjecture out of such a trifling investment of fact."[12]

The expectation that there should be many alien civilizations is due in part to the Drake equation, which mechanistically combines extremely uncertain factors. A group at the Future of Humanity Institute in Oxford has recast the equation to utilize realistic distributions of uncertainty. It found no reason to be confident that the galaxy or even the observable universe contains other civilizations. The bottom line: "This dissolves the Fermi paradox, and in doing so removes any need to invoke speculative mechanisms by which civilizations would inevitably fail to have observable effects upon the universe."[13]

David Brin has long cast a skeptical eye on some of the extravagant speculations in the field of xenology, the scientific study of alien life. He's a scientist, author of "hard" science fiction, and the winner of multiple Hugo and Nebula awards. Forty years ago, Brin wrote, "Xenology is not a field conducive to philosophical rigor. . . . Few important subjects are so data-poor, so subject to unwarranted and biased extrapolations—and so caught up in mankind's ultimate destiny—as is this one. Many prominent scientists have been drawn into a debate which has, with very few exceptions, fallen into a confusion of ill-matched arguments and non-sequitur." Ouch. He continues: "The quandary of the Great Silence gives the infant field of xenology its first traumatic struggle, between those who seek optimistic excuses for the apparent absence of sentient neighbors and those who enthusiastically accept the Silence as evidence for humanity's isolation in an open frontier." Brin notes that many of the scenarios that remain viable are depressing or dark, but he closes on an optimistic note: "It might turn out that the Great Silence is like that of a child's nursery, wherein adults speak softly, lest they disturb the infant's extravagant and colorful time of dreaming."[14]

Speaking of speaking, SETI involves listening. What about the other side of the coin? What are the implications of the fact that we're listening but not talking into space?

It's the astronomer's version of the single person's dilemma. Should you sit and wait for a call from that special person? Or do you make the call and risk the phone ringing endlessly—or worse, risk getting shot down? If many civilizations are technically capable of contacting Earth, but they're all or almost all listening rather than transmitting, then SETI will fail, and we'd have another answer to Fermi's question.

Our efforts to send messages into the cosmos have been meager and limited. NASA's Pioneer 10 and 11 spacecraft, launched in the early 1970s, carried identical plaques with pictorial messages intended for intelligent extraterrestrials. The plaque's designers—Drake, Sagan, and Sagan's first wife, Linda Salzman—were given three weeks to do their work. The plaque portrayed silhouettes of a man, a woman, and the spacecraft, a symbolic representation of a fundamental transition of hydrogen, and a pulsar map to triangulate the location of Earth.[15] Capturing humanity in one engraved image is already challenging, but almost inevitably, the plaque became a lightning rod for criticism when the image was released to the public. Objections included the opacity of some of the symbolism, implicit sexual bias in the human figures, and the fact that they were naked.[16] The plaques were the first two human-made objects to leave our solar system; both are currently more than a hundred times the distance from Earth to the Sun. It will be millions of years before either of them encounters any nearby stars.

A few years later, Sagan chaired a committee to design another artifact that would be sent into space as a message from our species. This time, it was a phonograph record carried aboard the Voyager 1 and 2 spacecraft. The Voyager record was intended to be a time capsule, and its cover included instructions on how to extract the analog information recorded in the grooves. The record included 115 images, sounds of nature, spoken greetings in 55 ancient and modern languages, and 31 musical selections.[17] When it comes to music to represent the planet, everyone gets to be a critic. The team mostly played it safe with folk songs, indigenous music, classical masterworks. Three selections of Johann Sebastian Bach were the most for any composer, but not enough for writer Lewis Thomas, who opined, "I would vote for Bach, all of Bach, streamed out into space, over and over again. We would be bragging of course, but . . . we can tell the harder truths later."[18] With popular music, the team members played it safe—a little jazz and a little blues. A selection by Chuck Berry was as edgy as they got. Frozen in the early 1970s, there was no disco, punk, or rap. The Beatles didn't make it since the company that held their song rights, EMI, displayed cosmic myopia by refusing to let their music be played "in perpetuity, across the known universe."[19] The team members took flack for using an analog LP, a technology that was already showing signs of becoming obsolete on Earth. They pushed back, noting that the gold-plated record was designed to last a hundred thousand years—something that can't be said of the more modern digital media of CDs and DVDs. The Voyager spacecraft are in interstellar space and they're still sending back data—so we can be fairly confident that no aliens have played the discs yet.

These efforts seem quixotic, like messages in a bottle tossed into the interstellar sea. If aliens intercepted the Voyager record, it would only be meaningful if they had ears, an atmosphere to carry sounds, and the sensibility to appreciate the patterns that we call music. Rather than a serious attempt to communicate, these artifacts are reflections of humanity, putting our best foot forward as we face an enigmatic universe.

The counterpoint to SETI is messaging extraterrestrial intelligence (METI). To organize the activity, METI International was founded in 2015 in San Francisco as a nonprofit research corporation. Long before the founding of METI, in 1974, radio astronomers sent a message from the 305-meter (1,000-foot) Arecibo radio telescope dish in Puerto Rico. The message, beamed out into space during a ceremony to mark the upgrading of the facility, consisted of 1,679 binary digits expressed as pulsed radio waves. The broadcast lasted three minutes. To decode the pictograms represented in the message, putative aliens would have to recognize 1,679 as the product of two prime numbers, arrange the digits in a two-dimensional grid, choose the correct one of the two possible arrangements, and realize that if on-off radio strength is mapped to light and dark (and you're an alien with eyes!), an image can be seen. The message will take twenty-five thousand years to get to its destination, the globular cluster M13, making this an extremely long-term experiment. More recently, in 2017, METI researchers sent a message to a closer target, the red dwarf Luyten's star, twelve light-years from Earth.

Contact with extraterrestrial intelligence would have profound consequences for the world. Even if it's unlikely to occur, policy makers need to be involved in deciding whether we should call attention to ourselves, who speaks for Earth, what message we should send and who decides, and how the communication would be managed.[20]

METI is controversial. Luminaries like Elon Musk and Stephen Hawking have said we shouldn't send messages since we have no way of knowing if aliens would be benign or malign. Instead of cute and lovable *E.T.*, we might get the planet-destroying aliens from *Independence Day*. When separate civilizations have met here on Earth, it has not usually gone well; think, for example, of the disease and slaughter that Europeans brought to the Americas. An online statement summarizes this perspective: "Intentionally signaling other civilizations in the Milky Way Galaxy raises concerns from all the people of Earth, about both the message and the consequences of contact. A worldwide scientific, political and humanitarian discussion must occur before any message is sent."[21]

I'm on the METI Advisory Council, an international group that discusses strategies for messaging and considers the societal implications of alien contact. Our group includes astronomers, philosophers, sociologists, and ethicists. The conversations are wide-ranging and invigorating, even if they do tend toward the esoteric. It might seem fanciful to take seriously communicating with aliens when we don't even know if there's any life beyond Earth, but our group views messaging as an optimistic way to affirm the role of humanity beyond our daily existence.

Chinese science fiction writer Liu Cixin has given voice to the legitimate fear of contact: "The universe is a dark forest. Every civilization is an armed hunter stalking through the trees like a ghost, gently pushing aside branches that block the path and trying to tread without sound. Even breathing is done with care. The hunter has to be careful, because everywhere in the forest are stealthy hunters like him. If he finds other life—another hunter, an angel or a demon, a delicate infant or a tottering old man, a fairy, or a demigod—there's only one thing he can do: open fire and eliminate them. In this forest, hell is other people. An eternal threat that any life that exposes its own existence will be swiftly wiped out. This is the picture of cosmic civilization. It's the explanation for the Fermi Paradox."[22]

We shiver and recoil from such a dark vision. A more optimistic reaction to intelligent aliens of unknown function and form comes from poet and writer Diane Ackerman. Her "Ode to the Alien" imagines that aliens will be no more strange or remarkable than we are. It finishes with, "My blood runs laps; I doubt yours does, but we share an abstract fever called thought, a common swelter of a sun. So, Beast, pause a moment, you are welcome here. I am life, and life loves life."[23]

IV
The Promise of Space Exploration

The litany is depressingly familiar: much of the world's arable land has been lost to erosion, pollution, or overuse. Fossil fuel supplies are finite. Mineral resources are being depleted ever more rapidly. Earth is being sullied by our waste. We're on an unsustainable trajectory.

The "box" is our planet, and we must think outside it. Dramatic progress in the space industry may offer a solution. Private companies like SpaceX and Blue Origin are rapidly lowering the cost of launching a kilogram into Earth orbit. When it drops by another factor of ten, space tourism will be viable, and companies will enact plans to harvest energy and minerals from places off Earth. This rapid innovation could allow us to transcend the resource bottlenecks we face on the home planet.

That's one argument and one way forward. But the flip side says space exploration is a wasteful extravagance. We're far better off investing money and energy in taking care of the home planet. Untrammeled space exploration will simply repeat the sins of the industrial age, bowing to the misguided calculus of capitalism and valuing economic growth over human well-being. Settlements in space will extend inequality into new realms. We've no right to exploit the Moon or Mars; our obligation to stewardship extends off Earth.

The tension between these two views of space exploration needs to be resolved because the activity is growing whether we like it or not. In this last set of chapters, I look at a road map of how we might ameliorate some of our problems with the judicious exploration of space.

21
Habitable Earth

*The connective tissue of the narrative of this book is habitability—
the likelihood that there are myriad havens where biology prospers,
contrasted with the fact that we're rendering our planet less habitable.
Habitability is a moving target. We imagine Earth to be an Eden
compared to the hostile environments of planets beyond our solar
system. Yet for most of its history, our planet would have been uninhabitable by humans. We live on a knife-edge where our capabilities
and impacts on the environment are increasing dramatically within a
human lifetime. Technology can be our salvation or the agent of our
destruction.*

*In this chapter, I'll take us on a whistle-stop tour of the history of
our planet, ending with its current state, which finds us on the cusp
of tilting into a state where we might struggle to survive.*

Hell is too gentle a word to describe it. For half a billion years,
Earth was molten as it released the heat from its formation. Even
as the crust formed, large areas of it were reliquefied by the impacts
of debris left over from the formation of the Sun and planets. The
atmosphere was a toxic cocktail of ammonia, methane, and carbon dioxide. Clouds of steam shrouded the surface as unimaginably large volumes of water were vaporized every time a chunk
of space debris slammed into the young ocean at hundreds of
thousands of miles per hour. If we explored exoplanets and happened on this hostile terrain, we'd shudder and move on, certain
that it was inhospitable to life. Yet in this harrowing place, simple
chemicals were combining to form complex molecules.[1] They were

concentrated into small containers made of fatty molecules. Over millions of years, random interactions led to some molecules growing into long carbon chains. Their bonds held the rudiments of information. In a slow transition that's still mysterious to scientists on present-day Earth, natural selection in favor of chemical complexity turned into biology, where simple cells survived by adaptation in a harsh environment.

This is probably the path taken by many earthlike worlds in the universe. Welcome to our newly habitable planet.

After more than 3 billion years of slow and steady biological evolution, an alien visiting Earth would have found no organism larger than the head of a pin. Starting 750 million years ago, the planet suffered a series of spasms. The first was from the breakup of a supercontinent that scientists call Rodinia and then from three extreme glaciations, each of which turned the planet into "Snowball Earth," with glaciers reaching nearly to the equator.[2] Earth was poised on a knife-edge, close to being locked in the deep freeze forever. Luckily for us, volcanoes belched enough carbon dioxide into the atmosphere to warm the climate. As the clamps came off, the evolution of plants and animals accelerated. The first animal was a humble sea sponge, but half a billion years ago, the seas were home to a proliferation of exotic creatures (as a result of the Cambrian explosion). Crude brains and central nervous systems appeared for the first time. Many of these evolutionary experiments didn't endure, but some plants and animals adapted and prospered, and over 100 million years they completed their takeover of the landmasses.[3]

If we checked in on Earth sixty-five million years ago, we'd find a planet reeling from two devastating events. First, a primeval range of volcanoes in central Asia created lava flows a thousand miles long. So much carbon dioxide was pumped into the air that the food web was disrupted and most large animals went extinct. Then a city-size meteorite plowed into the shallow sea off the coast of what is now Mexico. Tsunamis circled the globe and dust flung into the air chilled the climate. Acid rain poured from the sky and the global temperatures swung from frigid to scorching. Most surviving species of large animals were killed off. Dinosaurs no longer ruled the land and marine reptiles no longer ruled the oceans. Some small mammals survived. No bigger than shrews, they ventured into the evolutionary void created by the mass extinction event.[4] As the climate became more temperate, vast new swathes of fresh land were created. Around six million years ago, there lived a primate ancestor common to both humans and chimpanzees. Four million years later, the first members of the genus *Homo* were walking in the East African Rift Valley. Back then, the savannas supported wild dogs the size of grizzly bears, hell pigs with the bulk of a modern rhinoceros, and grazers to make an elephant look small. It's unclear if we would have survived in such an environment. Yet the stage was set for us. Our species, *Homo sapiens*, appeared around two hundred thousand years ago in East Africa.

By the time Earth emerged from the grip of the last glacial period, twelve thousand years ago, humans had radiated out of Africa and inhabited every ice-free region. They numbered five million, so their impact on the planet was small.[5] As the global temperature warmed, they gathered and settled in the fertile river valleys of Asia Minor and developed agriculture. For the first time, they didn't live entirely opportunistically. They planted crops, irrigated land, forged metal tools, and planned for a future beyond their lifetimes. They began to control the habitability of their environment. They started to selectively breed plants and animals; they were no longer hunter-gatherers. Around seven thousand years ago, humans founded the great civilizations of Mesopotamia, the Indus River Valley, and Egypt. Progress was interrupted by plagues and famine. Ironically, the invention of agriculture narrowed food diversity and the overall quality of nutrition declined.[6] And in an ominous sign for the future, settled communities became vulnerable to communicable diseases. Germs evolve thousands of times faster than people, so we will always be challenged in this type of evolutionary arms race. As people tried to preserve and expand their resources, conflicts became more common. Humanity was on a trajectory that would transform the planet.

"That's one small step for man, one giant leap for mankind."[7] The black-and-white video signal was grainy, the audio had an awkward lag, and the man was clumsy in his bulky spacesuit. But this first step on another world was as profound as when humans first stepped out of Africa fifty thousand years before.

It took two hundred thousand years for the human population to reach 1 billion in 1804, just over a hundred years to reach 2 billion in 1927, and just over thirty years to reach the third in 1960.[8] By 1969, when Neil Armstrong spoke those iconic words on the lunar surface, the world's human population had ballooned to 3.5 billion; it's now around 8 billion. It hasn't been a smooth road. Between the first arable crops and those first steps on the Moon, half a billion people died in pandemics and the same number in wars. Despite the success of the iconic Moon landings, numerous warning lights are flashing. The world's superpowers have a stockpile of sixty thousand nuclear weapons. The world economy is fueled in an unsustainable way by the remains of long-dead plants and animals. In the late 1960s, Earth is about to start a dramatic warming trend. A billion people are worried about their next meal.[9] In reflective moments, the "master" species of the planet poses a question with no clear answer: Will we survive our troubled adolescence as a species?

We're surrounded by exponential change. Exponential growth describes any quantity that increases at a rate proportional to its current size. Most people learned about exponential growth with the coronavirus; it's amazing that infections can scale up quickly from a few people to affect a whole population. We saw in the discussion of energy footprints that Earth has entered an era called the Anthropocene, where human activity is the dominant influence on climate and the environment. In the context of astrobiology, Earth can be considered a "hybrid" planet, where the energy flows are out of equilibrium (see chapter 18) and the biosphere cannot catch up.[10] Simply put, we're out of balance. The exponential changes have been called the Great Acceleration, which began in the mid-twentieth century. A snapshot of our planet reveals alarming trends.[11]

Steeply climbing socioeconomic indicators include world population, and specifically urban population, real gross domestic product, foreign direct investment, primary energy use, water use, transportation, telecommunications, international tourism, fertilizer consumption, and paper production. Added (and related) to these are a dozen Earth system trends: the greenhouse atmospheric gases carbon dioxide, methane, and nitrous oxide, stratospheric ozone, surface temperature, ocean acidification, marine fish capture and shrimp aquaculture, coastal nitrogen, domesticated land, tropical forest loss, and overall terrestrial biosphere degradation. Parts of the litany are familiar, but it's shocking to see it collected. The danger is real and proximate. Will Steffen, who coined the term "Great Acceleration," led a team that identified four out of nine planetary boundaries where we've pushed through the planet's "safe operating limits." Steffen observed, "Transgressing a boundary increases the risk that human activities could inadvertently drive the Earth system into a much less hospitable state, thereby damaging efforts to reduce poverty and leading to a deterioration of wellbeing in many parts of the world, including wealthy countries."[12]

There are inequities under the surface of these global trends. They're revealed when the trends are divided into wealthy nations, emerging economies, and the developing world. The lion's share of consumption lies with the wealthy, industrialized countries of the Organization for Economic Cooperation and Development. They have 18 percent of the world's population yet account for 74 percent of the gross domestic product. But emerging economies like China, India, and Brazil now show significant contributions to production.[13] Here's Steffen again: "By treating humans as a single,

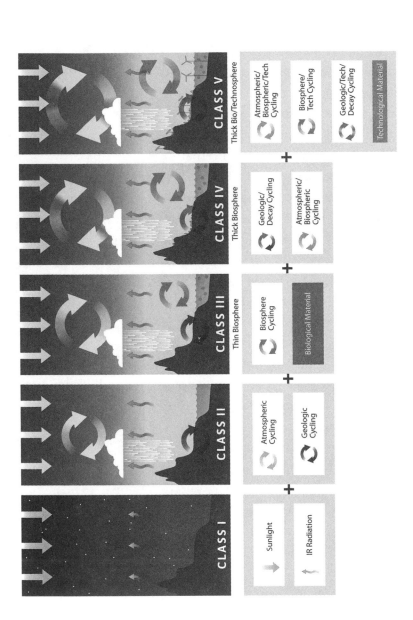

Figure 21.1

Energy flow for planets. Class I planets have no atmosphere, so solar radiation is reradiated into space. Class II planets have an atmosphere with some geological cycling. Class III planets have a thin atmosphere with biology part of the cycling of energy. Class IV planets have thicker atmospheres, and biology maintains energy equilibrium. Class V planets, as Earth is becoming, have technology that is driving the energy balance out of equilibrium. (Credit: A. Frank, A. Kleidon, and M. Alberti, in "Earth as a Habitable Planet," 2017, *Anthropocene*, volume 19, https://www.sciencedirect.com/science/article/pii/S2213305417300425#!)

monolithic whole, we ignore the fact that the Great Acceleration has, until very recently, been almost entirely driven by a small fraction of the human population, those in developed countries."[14] We should recognize the irony in the metaphor that a rising tide lifts all boats. With climate change, there's a literal rising tide that will inundate hundreds of millions of coastal poor in Asia, displace Pacific Islanders, and melt away the livelihoods of the Indigenous people of the taiga and Arctic.

The division into industrialized and developing countries hides the stark inequality of the modern world. Wealth and income inequality have increased dramatically since the 1980s.[15] The richest 10 percent of the world's population take home 52 percent of the income, while the poorest half earn just 8 percent. The gap is even wider for wealth. The richest 10 percent own 76 percent of the wealth while the poorest half own just 2 percent of the global total.[16] The economic growth behind the ascendancy of Europe was based on colonialism, the slave trade, and the dispossession of Indigenous and native populations. This dynamic has not disappeared in the modern era of economic globalization.[17]

Steffen has been speaking uncomfortable truths for decades, taking his data to places like the Davos Economic Forum, where the world's elites gather to orchestrate the future. He was trained as a chemist, and has numerous publications on climate science and sustainability. Steffen was executive director of the International Geosphere-Biosphere Program, a coordinating body of national environmental change organizations based in Stockholm, and has firsthand experience of the influence of politics on environmental policy. He was a member of the Australian Climate Commission when he was sacked, and the commission was disbanded, by an incoming government in 2013. "I think we were the first definitive action of the Abbott government," Steffen said. "They got rid of us and you could probably measure it in hours rather than days."[18]

The Great Acceleration started with the Industrial Revolution, and some growth has been positive and sustainable. The population was turbocharged by revolutions in agriculture and medicine that let more people eat and live longer. Telecommunications is a positive force; the wireless internet puts information in the hands of people in the developing world and lets them run their lives more efficiently. Even urbanization can be beneficial since city dwellers have smaller carbon footprints than those in rural areas. Fortunately, perhaps, we've "flattened the curve" on population. Predictions suggest a peak in thirty years at between ten and eleven billion. Other curves are bending, but the stabilization is limited by capacity; we're reaching limits of available land and fish stocks. Big gains in agriculture are behind us. A team of European scientists showed that eighteen out of twenty renewable resources have passed their peak production, including foods that feed the world—maize, wheat, rice, soya, fish, meat, milk, and eggs.[19]

I was in a small boat on a remote river in Borneo when I came face-to-face with our heavy footprint on the planet. I'd just seen the 2016 solar eclipse from a beach on Belitung Island, off the coast of Borneo, and now I was venturing into the rain forest to see orangutans in the wild. The moist air carried the scent of orchids and the calls of howler monkeys. It was as remote a place as I've ever been; there was no sign of civilization. The nearest small town was a hundred miles away. But as we rounded a curve in the river, the dense vegetation on one riverbank had been cleared, and we could see an endless vista of land that had been razed for palm oil plantations. The burning stubble sent a pall of smoke into the sky. At the next curve in the river, an eddy of water created a vortex that had trapped thousands of plastic bottles. Each one had been a drink for someone but now was destined to be a blight on the planet for centuries.

We're reaching the carrying capacity of our planet.[20] By 2050, all the available land will have been taken for agriculture, and ecological modeling suggests we will lose 40 percent of the biomass, 80 percent of large animals, and 90 percent of soil nutrients.

In addition, our industrial "debris" has reached a staggering scale. Pause a moment to let these numbers sink in. The amount of concrete we've generated over the past fifty years exceeds the weight of all humans. The amount of plastic we've produced is twice the weight of all animals on Earth combined. In 2020, the mass of human-created, artificial products—1.1 trillion metric tons—for the first time exceeded the entire biomass on Earth.[21]

The clearest way to understand where we've been and where we're going is to look at energy use. In the paper "Human Domination of the Biosphere," John Schramski and his coauthors lay it out clearly: "Earth is a chemical battery where, over evolutionary time with a trickle-charge of photosynthesis using solar energy, billions of tons of living biomass were stored in forests and other ecosystems and in vast reserves of fossil fuels. In just the last few hundred years, humans extracted exploitable energy from these living and fossilized biomass fuels to build the modern industrial-technological-informational economy, to grow our population to more than 7 billion, and to transform the biogeochemical cycles and biodiversity of the Earth." The laws of thermodynamics governing the trickle charge and rapid discharge of Earth's battery are absolute and universal. The coauthors end with an ominous warning: "With the rapid depletion of this chemical energy, Earth is shifting back toward the inhospitable equilibrium of outer space with fundamental ramifications for the biosphere and for humanity. Because there is no substitute or replacement energy for living biomass, the remaining distance from equilibrium that will be required to support human life is unknown."[22]

Here's how we got to be bigger than the biosphere. At rest, human metabolism requires about a hundred watts. We eat food to stay alive, and our bodies burn like a single light bulb (a bright old-style incandescent one) or flat-screen TV. Yet along with our biological metabolism, we must consider our extended metabolism: all the energy we burn with our daily activities. Humans before agriculture had an extended metabolism of three hundred watts. That's what it took to fuel a nomadic lifestyle of hunting and gathering. The population density was low—about one person for every ten square kilometers. Since the biological metabolism of the tropical savanna was thirty megawatts per square kilometer, humans only used 0.0001 percent of the savanna's energy capacity. We had a light footprint on the planet.

With the invention of agriculture, human energy requirements increased. Each person still burned like a single light bulb, but our "extended metabolism" increased to two thousand watts, thanks to the use of fire and beasts of burden. We started to concentrate on arable areas, and the population grew. Our energy use grew to 5 percent of the natural ecosystem metabolism. Fast-forward to the modern, industrial age. Taking the example of a densely populated European country like England, a person's extended metabolism has grown to eight thousand watts. The capacity of the natural environment, however, hasn't changed.[23] The modern European citizen operates at 200 percent of the metabolism of the ecosystem. It's unsustainable.

How big are we? There's a remarkable scaling relation in biology between the mass of an organism and its metabolic rate called Kleiber's law.[24] It spans from single-celled organisms up to elephants. A dog with a mass of ten kilograms (twenty-two pounds) burns about thirty watts. As we've seen, humans have a natural metabolism of a hundred watts. An African elephant with a mass of three thousand kilograms (nearly three tons) burns energy at a rate of five thousand watts. But modern North Americans have the average resource consumption of a thirty-ton primate. Each of us is like King Kong stomping on the planet!

Someone who knows about the limitless and finite is Ellen MacArthur. This British sailor bought her first boat with school lunch money, and at age twenty-four, she was the youngest person to complete a solo yacht voyage around the world. Four years later, she broke the speed record for sailing solo around the world. She said, "No experience in my life could have given me a better understanding of the word 'finite.' What we have out there is all we have. There is no more."[25] She knew, as we all now realize, that resources are finite and being rapidly consumed. We thought we had a limitless world. But really, we've just been sailing in a boat.

Our crowded and bruised planet spins through the blackness of space—a space that contains other worlds, large and small. In a diminished future, could we augment our finite supplies of energy and resources by harvesting from outer space? At the very least, space gives us a fresh perspective on the home planet.

"Viewed from the distance of the Moon, the astonishing thing about the earth, catching the breath, is that it is alive. The photographs show the dry, pounded surface of the moon in the foreground, dead as an old bone. Aloft, floating free beneath the moist, gleaming membrane of the bright blue sky, is the rising earth, the only exuberant thing in this part of the cosmos." This is the biologist Lewis Thomas, writing in his collection of essays called *The Lives of a Cell*.[26]

Astronauts of all nationalities have been struck by their view. from orbit of a fragile and seamless planet, with no boundaries or divisions. This profound cognitive shift has been called the "overview effect."[27] The space perspective renders parochial the rivalries and rifts that often consume us. The Apollo 8 image of earthrise in 1968 played a significant role in launching the modern environmental movement. Space has the power to inspire us. As we start to expand our footprint off Earth, could it also have the potential to solve some of our most pressing problems?

22
Space Boom

Before we can consider what benefits we might gain by expanding our exploration of space, it's worth considering how access to space might become easier. Until recently, space travel was difficult, dangerous, expensive, and the exclusive domain of rich governments. But private space companies are forging a novel economic model for space travel based on tourism, recreation, and eventually, the harvesting of resources off Earth. There are over two dozen space companies with ambitious plans. SpaceX and Blue Origin, companies headed by the two richest people in the world, are leading the charge.

In 2017, for the first time since Sputnik launched the space age in 1957, there were more commercial launches into Earth orbit than launches by governments.[1] Over the past decade, the price of launching into Earth orbit has dropped from ten thousand to a thousand dollars per kilo. Innovation and competition will drop that to a hundred dollars per kilo in fifteen years. Going into space will cost less than a three-week cruise.

What we're witnessing is analogous to the pioneering era of civil aviation. Back in the 1930s, flying in commercial airplanes was dangerous, expensive, and only for the elite. Now it's safer than driving, cheap, and easy enough that over four billion people flew in 2019.

In the same way, space travel promises to become routine. But the analogy with civil aviation isn't perfect. Commercial jets offer a relatively quick way to cross a continent or ocean. Space travel isn't about getting from A to B more efficiently. It's about taking people into an alien environment for tourism or adventure. There may come a time when spaceships regularly ply routes to the Moon and Mars, but it's decades in the future. Also, the regulation of space-flight is playing out differently from how civil aviation was regulated. Aviation industry leaders knew they needed government action to improve safety standards if the industry was to grow. In the United States, the Air Commerce Act was passed in 1926 and Federal Aviation Act in 1958, leading to the formation of the Federal Aviation Administration.[2] Air travel is amazingly safe, with a fatality rate of three per trillion kilometers traveled. Compare that to car fatalities, which are almost a thousand times higher per kilometer traveled.[3] The US government has kept a light hand on the rudder of the growing commercial space industry. Congress passed a law in 2004 that granted the private sector a "learning period" free of regulation. That period was extended three times and currently expires in October 2023.[4] There are regulations and safety rules, but people who participate in commercial spaceflight do so through "informed consent," meaning they know they're doing something that could easily kill them.

When we think of a "boom" in space travel, there's a second, uncomfortable meaning to the word. How dangerous is space travel? Not as dangerous as you might think.

Fewer than seven hundred people have even been into space.[5] Those lucky few have felt the disorienting effects of zero gravity and witnessed the sight of Earth's curved surface melting into the absolute blackness of space. Nineteen astronauts and cosmonauts haven't survived the experience. Another dozen died in training for spaceflight.[6] That's a death rate of 5 percent, which is safer than climbing above six thousand meters (twenty thousand feet) in the Himalayas, but more dangerous than hang gliding or base jumping.[7] Some astronauts have been in orbit five or six times, and each flight was like a game of Russian roulette where three guns were fired at their head, one of which had a bullet in a chamber. That's scary.

Statistics don't tell the full story; some of the fatal incidents are chilling to contemplate. The only people to die in space were three cosmonauts on their way back from the first visit to a space station in 1971. The Soyuz 11 crew was asphyxiated as its capsule depressurized before reentry. The two worst disasters involved NASA's space shuttle program. In 1986, *Challenger* blew up seventy-three seconds after takeoff due to the failure of a sealant ring on the booster rocket. The incident was particularly horrific because it was witnessed live by many schoolchildren since one of the crew was Christa McAuliffe, a specially selected teacher. In 2003, seven more astronauts died on *Columbia* when the shuttle broke apart on reentry into Earth's atmosphere. A piece of foam insulation broke off during launch and damaged a wing, sealing the mission's fate.[8] NASA was severely criticized for the loss of two of the five space shuttles, each due to the failure of a seemingly minor component. Another terrible incident was the loss of three astronauts training for the Apollo 1 mission in 1967 in a launchpad fire that almost derailed the Moon landings.

Being on the ground near a lot of high explosives can be hazardous as well. About a hundred people were killed in 1996 when a Chinese Long March rocket veered off course and crashed into a nearby village. More than that probably died in 1960, at the dawn of the space age, when a Soviet rocket exploded at the launchpad. The liquid fuel that the rocket carried was known as "devil's venom." It was toxic and corrosive in liquid form, and when it burned, it produced poisonous gas. People near the rocket were incinerated instantly; those farther away were burned to death in the fireball or poisoned by the toxic fuel vapor. The head of the rocket program, Chief Marshall Mitrofan Nedelin, was only identified by his partially melted Gold Star medal.[9] The worst rocket accident in history was hushed up by the Soviet Union for thirty years.

The infant private space industry doesn't have a spotless track record. Alongside SpaceX and Blue Origin, the triad of high-profile companies is rounded out by Virgin Galactic, headed by business magnate Richard Branson. In 2007, three people died after an explosion during an engine test for SpaceShipTwo, Virgin Galactic's suborbital spaceplane. In 2014, SpaceShipTwo suffered an in-flight breakup during a test flight. One pilot was killed, and another was seriously injured.[10] Blue Origin has had two dozen mostly successful launches since 2005 and sent its first passengers into space in 2021. SpaceX has flown and reflown the Falcon 9 rocket more than a hundred times with a high success rate. The company has also suffered a string of spectacular explosions, particularly of its experimental Starship design. Elon Musk is cavalier about these failures, believing that innovating quickly and breaking things is the way to stay ahead. After the fiery crash of a Starship rocket in February 2021, chief engineer John Insprucker was upbeat: "We had, again, another great flight up. We've just got to work on that landing a little bit." Musk was aggressively nonchalant: "This is a test program. We expect it to explode. It's weird if it doesn't explode, frankly. If you want to get the payload to orbit, you have to run things close to the edge."[11] Now that SpaceX is regularly launching people into orbit, the stakes and costs of failure are significantly larger.

Who's leading this charge into a zero-gravity future? The most prominent are two idiosyncratic billionaires who left their mark on the modern world even before they became space titans.

First, there's Musk, the man who tweeted that aliens built the Egyptian pyramids. He has a stark attitude toward failure: "My mentality is that of a samurai. I would rather commit seppuku than fail." He wishes he didn't have to eat so he could have more time for work. He's argued that humans live inside a video game: "There's a billion-to-one chance we're living in base reality." He's called for Mars to be nuked since thermonuclear weapons could warm the red planet as a prelude to human habitation.[12]

Even allowing for hyperbole and his thirst for media attention, Musk is one of the most eccentric and dynamic people in the business world. He was born in South Africa, and after gaining degrees in physics and economics, he dropped out of graduate school at Stanford after two days to launch an internet start-up. He joined the board of Tesla in 2004, and has steered that company from a minnow, with sales of twenty-five hundred cars in 2008, to the world's largest electric car company, with over a million sold by 2020.[13] Tesla's success has played a big part in nudging traditional car builders to make electric vehicles and so advance us toward a future based on renewable energy.

Musk founded SpaceX in 2002, using a hundred million dollars of his early fortune, and he's the CEO and chief engineer of the privately traded company. SpaceX started with three failed launches, but since then has racked up an impressive series of achievements. This was the first private company to reach orbit with a liquid-fueled rocket, place a commercial satellite in orbit, reuse an orbital rocket's first stage, put a spacecraft into a heliocentric orbit, send humans into orbit, and send astronauts to the International Space Station.[14] SpaceX's overall success rate for launches and landing the reusable boosters is above 95 percent.

Although it happened before he was born, Musk was inspired by the Moon landings: "What actually inspired me to create SpaceX was, I kept expecting that we would continue beyond Apollo 11, that we would have a base on the moon, that we would be sending people to Mars. And that by 2019 probably would be sending people to the moons of Jupiter. And I think actually if you ask most people in 1969, they would have expected that."[15] Musk was discouraged as he watched NASA flounder with the loss of two shuttles and dependence on expensive, expendable rockets. He tried to buy cast-off ICBMs from Russia, but the Russians gouged him on the price. In despair, he decided to go it alone: "I gotta try building a rocket company. I thought this was like almost certain to fail. In fact, I would not let anyone invest in the company in the beginning 'cause it was like, I can't take people's money. This is gonna fail. So I actually funded the whole company in the beginning myself. Not because I thought it would turn out well, but because I thought it would fail." He didn't fail. In 2020, SpaceX launched astronauts to the International Space Station—the first to get there from US soil in nearly a decade since the grounding of the space shuttle program. As icing on the cake, NASA let SpaceX use pad 39A at Kennedy Space Center, the fabled site of almost all the Apollo and space shuttle launches. Despite his irreverent streak, Musk respects history: "I can't believe we get to use this pad. An insane honor."[16]

Next, meet Jeff Bezos, a man who founded a company that sells four thousand items a minute and is the world's largest online marketplace. His second choice for a name for his Amazon company was Relentless. He's blunt with his employees as these rhetorical questions illustrate: "Why are you wasting my life?" and "Are you lazy or just incompetent?" and "I'm sorry, did I take my stupid pills today?" His philosophy for success and staying hungry is, "Be kind, be original, create more than you consume, and never, never, never let the universe smooth you into your surroundings. It remains Day 1."[17]

Bezos was a highflier from the start: a National Merit Scholar and high school valedictorian, and summa cum laude graduate of Princeton University with a double major in computer science and electrical engineering. Few others could have foreseen what lay in the future when he began an online bookstore called Amazon in 1994. But Bezos knew from the get-go that he wanted it to be the "everything" store. He takes a long view in all of his endeavors. Bezos sunk forty-two million dollars into a gargantuan mechanical clock being built inside a mountain in Texas. The "Clock of the Long Now" aims to be an antidote to the short-term thinking of modern society. It's designed to tell time accurately for ten thousand years.

Amazon has over a million employees and some jaw-dropping financials: a market capitalization of $1.6 trillion, and revenues of $400 billion—numbers that put the company in the top five in the world. Amazon has the largest market share, and near monopolistic clout, in selling books, providing cloud computing, and streaming video. The company was the subject of a blistering *New York Times* article that exposed its hypercompetitive, dog-eat-dog workplace culture.[18] Franklin Foer, who did a deep dive into Amazon's corporate culture, summed up its profound place in the world of commerce, and the power of Bezos himself, this way: "In the end, all that is admirable and fearsome about Amazon converges. Every item can be found on its site, which makes it the greatest shopping experience ever conceived. Every item can be found on its site, which means market power is dangerously concentrated in one company. . . . Bezos' company has become the shared national infrastructure; it shapes the future of the workplace with its robots; it will populate the skies with its drones; its website determines which industries thrive and which fall to the side. His investments in space travel may remake the heavens. The incapacity of the political system to ponder the problem of his power, let alone check it, guarantees his Long Now. Bezos is fixated on the distance because he knows it belongs to him."[19]

Bezos sells everything to everyone, but he's always dreamed of space. In his high school valedictorian speech, he conjured up a vision of hotels, amusement parks, and human colonies in orbit. He was inspired by the populist vision of moving into space espoused by Gerard K. O'Neil.[20] He has a geeky obsession with *Star Trek* in all of its incarnations and has even made his physical appearance converge with that of his fictional hero, Jean Luc Picard. Bezos founded Blue Origin in 2000, and the company has pursued an incremental approach to orbital flight—incremental and obsessive. The company motto is *Gradatim Ferociter*, Latin for "Step by Step, Ferociously." Bezos has kept Blue Origin going by selling a billion dollars of his Amazon equity every year. The company built a spaceport in West Texas, and it now has thirty-five hundred employees, a third the size of the SpaceX workforce. Bezos stepped down as CEO of Amazon in 2021 to devote himself more fully to his vision of our off-Earth future.

In the financial stratosphere that Bezos and Musk occupy, fortunes can be gained or lost quickly. Bezos added $13 billion to his fortune in fifteen minutes when Amazon beat fourth-quarter earnings estimates in 2019, and Musk lost $14 billion in four days after a sell-off in tech stocks in 2021. As for their rivalry to exploit and explore space, while it might look like Blue Origin is the plodding tortoise and SpaceX is the speedy hare, both companies have radically altered the economics of space travel with their reusable rockets. In July 2021, Bezos gained major bragging rights on Musk by being the first to fly his rocket into space. In turn, however, he had been beaten by ten days with Branson's ten-minute, zero-gravity joyride to the edge of space. Both were suborbital flights, so bragging rights for the first tech mogul to reach Earth orbit are still up for grabs. Bezos blended science with his science fiction obsession by publicly treating William Shatner (aka Captain Kirk on *Star Trek*) to a suborbital flight in October 2021.

Bezos and Musk agree on one thing: the key to bringing down the costs of launches to Earth orbit is reusability. The ruinous economics of NASA's space ventures boils down to the inefficiency of nonreusable space hardware. Here's Bezos: "Reusability is essential because you can never lower the costs to a sufficient degree if you throw the hardware away. That hardware is just too beautiful. First of all, it's just painful. You get this great feeling when looking at a piece of aerospace-grade hardware. It's so beautiful and so precise. To use it once and throw it away is a kind of crime."[21] Musk makes a similar point with an analogy: "It would be as though if, in the old days, ships were not reusable. The cost of an ocean voyage would be tremendous. And you'd need to have a second ship towed behind you just for the return journey. You can imagine if airplanes were not reusable. Nobody would fly 'cause each airliner costs a couple hundred million dollars. And people do not wanna pay that for a single journey. So, this is why full and rapid reusability is the holy grail of access to space and is a fundamental step towards it, without which we cannot become a multi-planet species."[22]

Branson is more than an afterthought—but he inevitably finds himself in the wake of the powerful progress being made by SpaceX and Blue Origin. He shares some traits with both Musk and Bezos.

Figure 22.1

The key to reducing the cost of launching to Earth orbit is reusability. Here, two SpaceX Falcon Heavy side boosters land after a successful flight, after which they can be refueled and reused. (Credit: SpaceX, 2018, https://commons.wikimedia.org/wiki/File:Falcon_Heavy_Side_Boosters_landing_on_LZ1_and_LZ2_-_2018_(25254688767).jpg.)

The titles of his two autobiographies convey his scurrilous streak: *Losing My Virginity* and *Screw It, Let's Do It*.[23] Like Musk, he has a self-aggrandizing streak and thirst for publicity. He transformed himself as he became an adult since he suffered from ADHD as well as dyslexia and was painfully shy as a child. He's made cameo appearances on many TV shows and a handful of films. Like Bezos, Branson started modestly, with a mail-order record company he set up at the age of twenty, only to diversify into a plethora of products.[24] At various times you could purchase Virgin brides and Virgin condoms, and you could drink Virgin vodka and Virgin cola. Branson wept when he had to sell Virgin Records to keep his new airline afloat in 1992. He launched his space tourism company, Virgin Galactic, in 2004. His genius move was hiring Burt Rutan, the premier aircraft designer of the past fifty years.[25] Virgin Galactic has had a bumpy road, with a fatal accident and benchmarks reached long after Branson's predictions. Yet the company has taken in over a hundred million dollars in deposits for its suborbital flights featuring six minutes of weightlessness, and like Musk and Bezos, Branson has invested heavily from his fortune to ensure its eventual success.

Where does all of this entrepreneurial activity leave NASA?

It's been argued that the space agency is so risk averse, and puts such a high value on human life, that it has become incapable of innovating.[26] This is the antithesis of the Musk credo of "develop fast and break things." NASA has struggled to replace the space shuttle with a new capability to lift heavy payloads into Earth orbit. After proceeding with difficulty for five years, the Constellation program was canceled. Its replacement, the Space Launch System, has cost $20 billion so far, and is costing $2.5 billion per year with no crewed launch yet scheduled. Nevertheless, NASA has recently become savvier. It's partnering with the private sector with a series of multibillion-dollar contracts to resupply the International Space Station. Both sides benefit. Companies like SpaceX get taxpayers to subsidize their substantial development costs, and the near-geriatric space agency learns to be nimbler.[27]

Space is booming. Branson, Musk, and Bezos get all the attention, but a vast majority of launches are still done by the major superpowers. The United States and China each had over fifty orbital launches in 2021, and there were more successful launches than any year in the history of the space program since Sputnik.[28]

23

How to Get to Space

"Earth is the cradle of mankind, but mankind cannot stay in the cradle forever."[1] So said a Russian mathematics teacher named Konstantin Tsiolkovsky in 1903. Inspired by Jules Verne, Tsiolkovsky thought populating space would lead to the perfection of the human species. As we'll see, he was a visionary in astronautics.

Not only did Tsiolkovsky develop the concept behind space-going rockets—which until now have been the only way to lift people and cargo into space. He also envisaged another technology, which holds the promise of making future space exploration much easier, safer, and cheaper.

If we step back from the churn and tumult of the commercial space race, we can see a progression with four distinct phases. It's useful to compare that progression with the development of a central tool of modern life: the internet.[2]

The first phase in the development of space travel belongs to the visionaries and pioneers. In 1903, Tsiolkovsky wrote a book called *Exploration of Outer Space by Means of Rocket Devices* that presented the equation at the heart of rocketry, and laid out how to use multi-stage rockets fueled by liquid oxygen and liquid hydrogen to reach Earth orbit. In 1912, US-based inventor Robert Goddard effectively developed the same equation. In the 1910s, he experimented with solid-fuel rockets—and in 1926, he launched the world's first liquid-fueled rocket. Although the rocket only traveled for less than three seconds, ascended to just 12 meters (40 feet), and traveled a total of 56 meters (184 feet) over a frozen cabbage field on his Aunt Effie's farm, it was the precursor to the mighty Saturn V rocket that would

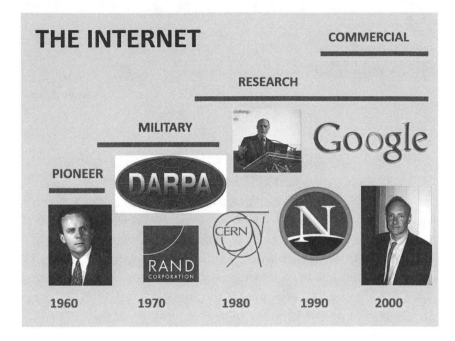

Figure 23.1
The internet had a pioneering phase where people envisaged the future, then a phase of incubation by military agencies and contractors, and then a phase of development by research institutes and universities, followed by the current phase of explosive growth of the commercial sector. (Credit: C. Impey.)

carry astronauts to the Moon. For the internet, the analogous first phase starts with Claude Shannon's foundational paper on information theory in 1948, and a 1960 paper titled "Man-Computer Symbiosis" by J. C. R. Licklider that envisaged a worldwide computing network and data in the cloud at a time when the few computers that existed were the size of a family house.

In the second phase, the nascent activity was incubated by the military-industrial complex. At the end of the Second World War, German aerospace engineer Wernher von Braun surrendered to US troops, and his work on the lethal V2 rocket was redirected into the development of ballistic missiles for the US Army at the height of the Cold War. The United States' use of von Braun's talents was expedient rather than principled; his Nazi connections were scrubbed to give him security clearance. NASA was established in 1958, and to his credit, President Dwight Eisenhower resisted the entreaties of his generals to make it a military agency. NASA's charter put it under civilian control, with transparent reporting and budgets allocated by Congress. Von Braun became the first director of the Marshall Space Flight Center, devoting his considerable talents to the Saturn V rocket and the goal of landing on the Moon. The internet also began with funding from the US military, under the Department of Defense's Advanced Research Projects Agency. ARPANET, the first computer network, was launched in 1969. Email, the killer application of the internet, developed around the same time. For nearly a decade, these tools were only available to people in the US military along with a few universities and research institutes.

The third phase saw the activity move firmly into civilian life, primarily for research. NASA's budget peaked at 4.5 percent of the federal budget in 1967 to achieve the herculean feat of getting to the Moon. Since then, it only briefly exceeded 1 percent at the peak of space shuttle launches. Recently, it has settled to around 0.5 percent of the federal purse.[3] Post-Apollo, NASA's human spaceflight endeavors concentrated on the space shuttle and International Space Station. There were 135 shuttle launches over thirty years, and while most launch components were reusable, the huge external fuel tank wasn't. Two space shuttles were lost catastrophically, and it was an extremely expensive conveyance. The military put payloads on several dozen flights, but eventually grew frustrated by costs and delays, so made separate arrangements to get to orbit. The International Space Station is a high-profile collaboration with contributions from five different country's space agencies. It demonstrates our ability to live and work in space; it's been continuously occupied for over twenty years. But it has cost over $150 billion. Neither facility attracted the level of interest from academic and industrial researchers that NASA had hoped for.

NASA operates in a political landscape that changes with every election cycle, compromising its ability to do long-term, strategic planning. Similarly, with the internet, government investment aimed at supporting research fueled development in the 1980s. The National Science Foundation funded a series of supercomputer centers and the creation of NSFNET, a high-speed successor to ARPANET. University computer scientists developed rules to allow computers to communicate over a network—transfer control protocol / internet protocol—and the naming system that is the internet's version of a phone book, the domain name system. Tim Berners-Lee, a physicist at CERN, developed hypertext markup language, the language to share and read documents over the internet, and a student at the University of Illinois, Marc Andreessen, invented the first web browser.[4]

What happened next? With the internet, all the tools were in place by the early 1990s for seamlessly sending data—text, images, and videos—anywhere in the world and displaying them on personal computers. In 1995, NSFNET was decommissioned and restrictions were removed on the use of the internet for commerce. What followed was an unprecedented surge in economic activity, as culture, commerce, and technology were transformed by the internet. The internet accounted for 1 percent of the information flowing through telecommunications networks in 1993, and that grew to 97 percent in 2007.[5] With one hiatus after the "dot-com bubble" in 2000, the internet economy has grown to over $2 trillion, and two of the four companies with a market capitalization above $1 trillion do their business primarily on the internet—Google and Amazon.[6] The space industry is booming too. While not at the level of the internet economy, it is predicted to triple to $1.4 trillion by the end of the decade.[7] The new players are disrupters in a good sense, creating new business models and questioning premises that have underpinned government investments in space travel.

Where the comparison with the internet breaks down is in the way technology scales and how that affects economics. The growth

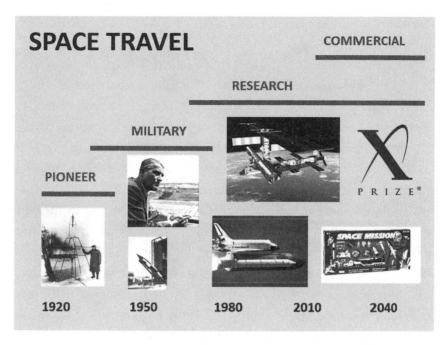

SPACE TRAVEL

COMMERCIAL

RESEARCH

MILITARY

PIONEER

X PRIZE®

1920   1950   1980   2010   2040

Figure 23.2
Space travel started with visionaries like Robert Goddard, was developed by building on military rockets, and then had a long phase of steady growth under NASA, followed by the current transition to a vibrant commercial space industry. (Credit: C. Impey.)

of the internet has been driven by exponential increases in information technology to deliver data. The most famous example is Moore's law, the doubling of the number of transistors in an integrated circuit every two years. Another illustration is the doubling of telecommunication network bandwidth every eighteen months. The world's computing ecosystem is growing exponentially, thereby supporting the delivery of ever-increasing amounts of information over the internet.[8]

By contrast, space travel is subject to the obdurate and immutable laws of physics, which presently hold it back from the more widespread availability enjoyed by the internet. Low Earth orbit is only 400 kilometers (250 miles), half a day's drive straight up. How hard could it be to get there? Very hard; gravity is a bitch. A warning to readers who have made it this far in a book about exoplanets: the next section really is rocket science!

We return to Tsiolkovsky and his equation. The man who derived this core principle of rocketry was born in a small town outside Moscow in the waning years of czarist Russia. At the age of ten, he lost most of his hearing after a bout of scarlet fever, and grew up isolated and homeschooled. He lost a son to suicide and many of his accumulated papers were destroyed in a flood. Living in a rural log cabin and teaching math in a provincial high school, Tsiolkovsky was isolated from academia, and his ideas were met with indifference or scorn, yet he was a true pioneer. He built a centrifuge and experimented with gravitational effects using chickens, built the first wind tunnel in the world, and was the first to propose the idea of a fully metal aircraft. Tsiolkovsky's 1903 equation is the foundation of astronautics. He also proposed multistage rockets, and the use of liquid hydrogen and liquid oxygen for rocket fuel, even though neither was yet available.

At first glance, a rocket seems magical. It shudders on the launchpad, billowing smoke and spewing fire, and slowly, majestically, it rises. How does it accelerate upward when there's nothing to push against?

The rapidly expanding exhaust gas is pushing upward on the inside of the rocket engine, propelling it upward. If there were nowhere for the gas to escape, its expansion would blow the engine apart, sending pieces of the rocket in all directions. But the gas can escape through the nozzle at the bottom, and as it does, the rocket accelerates in upward. Rockets are momentum machines. The upward momentum (mass times velocity) gained by the rocket is equal to the increased momentum of the gas escaping downward. Newton worked out the math for this momentum exchange in 1687, and in 1903, Tsiolkovsky applied it to rockets. Let's hear it from someone with firsthand experience, space shuttle chief engineer Don Pettit: "The rocket equation contains three variables. Given any two of these, the third becomes cast in stone. Hoping, wishing, or tantrums cannot alter this result. Although a momentum balance, these variables can be recast as energies. They are energy expenditure against gravity (change in rocket velocity), the energy available in your rocket propellant (exhaust velocity), and the propellant mass fraction (how much of the rocket mass is fuel)."[9]

The problem with getting a rocket into orbit is that it must accelerate all of its fuel from zero velocity on the launchpad. If you want to lift a heavier payload, you need more fuel. Then you have to accelerate that fuel, though, so you add more fuel, but then you have to accelerate that additional fuel, and on and on it goes. The "tyranny" of the rocket equation is that the final velocity of the rocket increases slowly as you add more and more fuel. Multiple motors don't help because they just allow you to burn fuel faster at the same exhaust velocity. You don't need any less fuel. Rockets must burn their fuel at a wild rate to create enough exhaust momentum to launch a payload. The space shuttle got through 2.3 million liters (500,000 gallons) of liquid propellant in just eight minutes—two million times faster than you burn gas driving around town in a car! Air resistance is another obstacle, particularly at low altitudes, and it's not accounted for in the rocket equation. As a result, rockets end up being mostly fuel, some metal, and a modest payload. The giant Saturn V rocket that sent astronauts to the Moon was 85 percent fuel, 13 percent rocket parts, and just 2 percent for the Moon-bound spacecraft with three astronauts inside. If Earth were 50 percent more massive, the diminishing returns of the rocket equation would make getting to orbit impossible with chemical fuel. Throughout the galaxy, there are more super-Earths than Earths, and any civilizations on these planets would be trapped unless they developed a rocket driven by nuclear fission, nuclear fusion, or matter-antimatter annihilation.

In this context, the gains made by SpaceX with mostly reusable rockets are impressive. The progression from Falcon 1 in 2010 to

Falcon Heavy in 2020, lowering the cost from ten thousand to a thousand dollars per kilogram, is a halving of the cost every three years. These companies have achieved efficiencies, but they're still working within the limitations of rocket physics. They're also limited by using chemical energy since rockets work the same way now that they did fifty years ago by combusting fuel and an oxidizer. Hydrogen burning with oxygen is the most energetic combination available for rockets. Chemistry is unable to yield any more energy.

Is there any way to be liberated from the tyranny of the rocket equation entirely? Yes! It's been called a sky hook, heavenly ladder, cosmic funicular, or prosaically, space elevator.

A space elevator is like a rope trick. The cable stretches up from Earth's surface with no apparent means of support. Cable cars carrying people and supplies move up and down using solar power. The space elevator works due to physics, and the weight of the cable is balanced by centrifugal force from Earth's rotation. Tsiolkovsky was the first to envisage this means of being liberated from Earth. He was inspired by the newly constructed Eiffel Tower in 1895, and speculated about building a tower reaching from Earth's surface to the height of a geostationary orbit. This is a circular orbit 35,785 kilometers (22,236 miles) above Earth's equator—around 5.5 times the equatorial radius of Earth. A satellite in this orbit revolves once around the planet in the same amount of time as Earth completes one rotation. As a result, the satellite appears stationary in the sky to an observer on the ground. Here's how Tsiolkovsky imagined it: "On the tower, as one climbed higher and higher up it, gravity would decrease gradually; and if it were constructed on Earth's equator and, therefore, rapidly rotated together with Earth, the gravitation would disappear not only because of the distance from the center of the planet, but also from the centrifugal force that is increasing proportionately to that distance. The gravitational force drops but the centrifugal force operating in the reverse direction increases. On Earth the gravity is finally eliminated at the top of the tower, at an elevation of 5.5 radii of Earth."[10]

The problem with Tsiolkovsky's concept is that a building is subject to compression. At some height, a tower cannot support its own weight. With graphite composite materials, a tower might be built 40 kilometers (25 miles) high, tapering from a 6-kilometer-wide (3.8-mile-wide) base. That's fifty times higher than the tallest skyscrapers, but only about one-thousandth the height of a geostationary orbit. A better concept was articulated by Arthur C. Clarke, first in science fiction in 1979, and then in a technical paper in 1981.[11] A long cable tethered to Earth's surface extends into space with its center of mass at a geostationary altitude. To keep the cable from tumbling to the ground, it must extend far beyond a geostationary altitude, where it's tethered by a large counterbalance mass. Gravity pulling the cable down is balanced by centrifugal force from Earth's rotation pulling it up. The space elevator is based on tension, not compression. The cable is essentially in orbit around Earth. Solar-powered lifters can carry payloads up and down the cable. When he wrote his article, Clarke knew there was no known material of sufficient tensile strength to build such a cable. He remarked, "The Space Elevator will be built about fifty years after everyone stops laughing."[12]

Well, nobody's laughing anymore. At a 1999 conference hosted by NASA's Advanced Projects Office, scientists and engineers rolled up their sleeves and tried to come up with a viable design. The organizer, David Smitherman, pointed out that "this is no longer science fiction. We came out of the workshop saying we may very well be able to do this." There are many technical challenges to building a space elevator, but finding materials of high enough tensile strength is the main one. Carbon nanotubes are light and a hundred times stronger than steel, and already have industrial applications. Even a single atomic-level flaw can cut their strength dramatically, so other materials under consideration are diamond nanothreads and single-crystal graphene. Here's Smitherman again: "What the workshop found was there are real materials in laboratories today that may be strong enough to construct this type of system."[13]

The past decade has seen increasingly sophisticated design studies, encapsulated in a 2019 report by the International Academy of Astronautics. The summary was that there are no technical "show-stoppers," and that real development is coming soon.[14] The practical details are being worked out, including how to prevent instabilities in such a long cable and what might be used as the counterweight. One wild and only mildly implausible idea is to capture a near-Earth asteroid for this purpose.[15] There are other serious considerations. For example, how might engineers design and power the "lifters"? Even at speeds of several hundred miles per hour, it would take a week to reach the geosynchronous altitude. A space elevator would represent a navigational hazard for aircraft and spacecraft, and have to be sturdy enough to survive occasional impacts from space debris. Finally, an iconic structure like a space elevator would be an attractive target for terrorists, so it would need associated defense systems.

A space elevator would dramatically change the economics of getting to space—no rocket fuel, no tyrannical equation, and no issue of reusability. SpaceX's Falcon Heavy has reduced the cost to low Earth orbit to around a thousand dollars per kilogram, which translates to about a hundred thousand dollars for an adult with some baggage. With a space elevator, that comes down to a few dollars per kilogram, or a few hundred dollars for the weeklong ride into space.[16] A dramatic cost reduction like that will open space for a new level of activity. The fact that the cable must extend far beyond a geostationary altitude enables the exploration of deep space. At the geostationary height of thirty-six thousand kilometers, an object released from the cable is in Earth orbit. At fifty-three thousand (thirty-three thousand miles), the object would be at Earth's escape velocity. If it could be extended to eighty-nine thousand kilometers (fifty-five thousand miles), an object could be sent to the outer solar system. The mind boggles to think of a cable this long, stretching a quarter of the way to the Moon.[17]

A huge benefit of a space elevator is that it can be environmentally friendly. As a recent paper from the International Space Elevator Consortium frames it as follows: "Space Elevators are the Green Road to Space—they enable humanity's most important missions by moving massive tonnage to high Earth orbit and beyond. They accomplish this safely, routinely, inexpensively, daily and they are environmentally neutral."[18] A space elevator avoids the need for wasteful and expensive rocket fuel. It's permanent transportation infrastructure with a zero-carbon footprint. It can also enable the construction of arrays for solar power in orbit, avoiding land use on Earth and moving us toward a sustainable energy future. As another side benefit, a space elevator could safely and cheaply move millions of tons of nuclear or toxic waste into orbit. If a space elevator could be used to dispose of hazardous material in deep space, it would solve one of our most vexing environmental problems.

If nobody's laughing anymore, the question is, When and for how much? Bradley Edwards, a physicist at Los Alamos National Laboratory, was commissioned by NASA to conduct a design study and deliver a cost estimate. He's also written several books on the topic.[19] Estimates range from five to ten billion dollars, in line with analogous transportation megaprojects like bridges, tunnels, and high-speed rail links. For example, the Channel Tunnel that connects England and France cost twenty billion in current dollars. The cost of a space elevator is far less than the previous means the United States had to reach Earth orbit: the space shuttle. The time frame for perfecting suitable cable material and working through the engineering challenges is probably twenty to thirty years. If one country or corporation built a space elevator, they'd have an enormous strategic advantage and could potentially control all space activities.[20] Fortunes will be made and lost in the brave new world off Earth.

24

Going to the Moon and Mars

We can see the outlines of a shining vision, with new opportunities for tourists and entrepreneurs—and perhaps humanity. Science fiction has long presented just about every aspect of this vision to us. But in the coming decades, with reusable spacecraft, economies of scale, and perhaps even a space elevator, it could become reality. The Moon and Mars beckon.

This is just the time when we should take a breath, pause, and consider the ethics of the continued expansion of space exploration. Could this activity also be for the benefit of humanity and the planet? Should we go back to the Moon and on to Mars?

Let's start with some of the legitimate criticisms about our ventures off Earth that should be addressed. Space travel is an expensive sideshow that we can ill afford given the problems on the home planet. The new commercial activity is billionaires with fancy toys driven by their egos. We should spend our energy and money taking better care of the terrestrial environment. Space travel will only ever be for the wealthy, so it will exacerbate existing inequalities. Humans have no right to exploit space resources with unforeseen consequences. Expanding our footprint beyond Earth will just replay the colonial and acquisitive history of the Western world in a new arena as well as on a bigger scale. There's so little law and regulation that applies to space that bad actors and unethical companies will face no constraints on their behavior (more on that in the next chapter).

Unfortunately, there's already one example where inattention, sloppy practices, and a lack of regulation are creating headaches for future space explorers: space junk.

Space junk is orbital debris—the detritus of our activity in space. It consists of chunks of metal, plastic, ceramic, and even flecks of paint that no longer serve a useful purpose. The sources of this debris are nonfunctional spacecraft, abandoned launch vehicles, cast-off materials from space missions, and fragmentation debris. There are twenty-three thousand pieces of debris larger than a softball orbiting Earth, tracked by the US Department of Defense's space surveillance network. Estimated populations of smaller pieces are half a million the size of a marble or larger, and a hundred million millimeter or larger. The problem is that they're all moving at extremely high speeds, up to 28,000 kilometers per hour (17,500 mph), and even a tiny bit of paint can damage a spacecraft at that speed. In the past, space shuttle windows have had to be replaced because of damage from paint flecks. It makes for quite a mental image: nuts and bolts hurtling through space ten times faster than a bullet. About once a year, astronauts on the International Space Station huddle in the sheltered central hub of the facility after getting a warning that debris is nearby.[1]

The problem is getting worse. As more satellites and spacecraft are launched, and more obsolete hardware accumulates in orbit, the probability of collisions increases. In 2007, the Chinese government intentionally destroyed one of its old weather satellites, adding thirty-five hundred chunks of large, trackable debris and many more small bits of debris to Earth orbit. Russia performed a similarly antisocial act in 2021. In 2009, a defunct Russian spacecraft slammed into an Iridium satellite, contributing twenty-three hundred new hunks of large space junk and many more small chunks. Commercial space companies like SpaceX are planning to launch tens of thousands of satellites in the next decade to facilitate wireless internet in parts of the world that have no coverage. Even before these plans were announced, it was predicted that collisions can cause cascading collisions, exponentially increasing the number and density of small pieces, and potentially rendering low Earth orbit completely unusable. This dire prospect is called the Kessler syndrome.[2]

We're facing a slow-motion catastrophe, not like the 2013 movie *Gravity*, where the opening scene showed low Earth orbit rapidly filling up with debris after a missile strike. Brian Weeden of the Secure World Foundation says, "In the movie *Gravity*, orbital debris was portrayed as sort of a nuclear chain reaction. The reality is the opposite. It's like climate change, a long, relatively slow accumulation of stuff over decades or longer that results in a really big negative impact down the road."[3] The problem has an ominous overtone because space is the fifth domain of war, alongside land, sea, air, and cyberspace. World powers are arming themselves to take out each other's satellites, offensively or defensively. It is going to grow increasingly difficult for a country to tell why their satellite went down or fell silent. Was it a collision with debris, space "weather," or a hostile action?

There's no international treaty governing space debris. A UN committee published voluntary guidelines in 2007. Mitigation strategies do exist, but governments and private companies have dragged their feet. Large satellites are now built with mechanisms for deorbiting, and some are in low enough orbits that drag in the thin upper atmosphere does the job in five to ten years. Technologies under consideration include lasers, nets, magnets, and harpoons. In 2019, ESA awarded the first contract for space cleanup. Yet the mission, called ClearSpace-1, won't launch until 2025. Earth orbit has become a new "tragedy of the commons," where we ruin something because we all profit from exploiting it and can't exclude others from doing the same. We've overgrazed public lands, overfished the oceans, and polluted the atmosphere, and we're in danger of ruining the sheath of space above our heads.[4]

"If you are expecting space ethics to tell you that space exploration is the greatest thing ever and that we should plunge ahead at all deliberate speed, then you may be in for a disappointment.

You are also in for a disappointment if you are expecting space ethics to validate calls to renounce space exploration and to accept our
terrestrial horizons." So wrote James Schwartz and Tony Mulligan,
two philosophers who specialize in the ethics of space exploration
and space policy.[5] How we deal with space debris is a test case, but
there will be many decisions to make as we explore the "new frontier." I've used quotes here because to boldly call space a frontier,
as something to "conquer," can bring up negative connotations. In
1893, historian Frederick Jackson Turner put forward his "Frontier
Thesis," in which he suggested that US democracy was created by
the conquest of the frontier, the wild space outside the newly civi-
lized towns. Anthropologist Lisa Messeri notes that Turner "mentioned economic pitfalls, specifically currency inflation, and in a
footnoted aside he acknowledged the gamblers and desperado
'scum,' also a product of the frontier."[6]

Space is a blank slate. We can write our future there and try to avoid
the mistakes that tarnish the legacy of many civilizations on Earth.

Over a hundred astronomers cosigned a white paper submitted
to the Planetary Science and Astrobiology Decadal Survey, a guiding document for research in the 2020s. They argued that "it is
crucial that the planetary science community, with community
input, take the opportunity before un-crewed and crewed exploration of other worlds to think ecologically—and seek to equitably
address the consequences of our presence on these other worlds."[7]
The private space companies are moving quickly, so there is some
urgency to develop this new field, as framed by this thoughtful
recent commentary: "Space ethics must embrace stewardship of
the space environment, the human rights of those endeavoring to
extend civilization into space, the rule of law, and how the benefits
of space can broadly benefit humanity while particularly motivating and rewarding those who risk, dare, invent, and invest."[8]

The field of space ethics is young and still evolving. But it's clear
what its role should be.[9] It should identify principles for arriving

at rational compromises between the different stakeholders in space. For example, should the Moon and Mars be protected in the same way national parks are, or should companies be able to use their land and resources without constraint?[10] Space ethics isn't for or against space exploration. It's a tool for identifying and critiquing assumptions that space advocates and space skeptics often fail to realize they're making. Not only can space ethics help us figure what's worth doing in space but it can also help us figure out the best way to do those things. For instance, if a goal of space resource exploitation is to improve human well-being, which methods achieve that goal with the least collateral damage to the space environment? Finally, space provides valuable perspective. Until now, activity in space has been dominated by US technocrats with a Western, Caucasian lineage. As we embark on multigenerational projects in space, we can aspire to represent the whole of humanity, examining a broad spectrum of human cultures and value systems. The practitioners of space ethics are primarily academic philosophers. It's ironic that they're mostly Western and Caucasian, but they do bring a welcome rigor to the subject.

There's a clear argument to be made that space isn't an expensive indulgence. If the satellites that whirl over our heads were to suddenly disappear, modern life would grind to a halt and there would be an economic catastrophe. We'd lose all global communications and media, GPS systems, weather forecasting, and the monitoring of critical resources like crops and water. Retreating fully to Earth isn't an option.

At this early stage of the endeavor, space is a blank canvas on which we paint our dreams, hopes, and aspirations. Strikingly diverse visions are on display. Journalist Adam Mann summed it up in an article on how we might treat other planets: "Space still exists mainly in our imaginations, and we all imagine it differently. At one extreme, there's the apocalyptic vision painted by Elon Musk, who wants to back up civilization on Mars in case of a catastrophe on our world. A more optimistic view comes from the 'Star Trek' franchise, which has shown humanity coming together in the spirit of science and exploration to discover strange new worlds."[11]

As we contemplate routine travel to the Moon and Mars, it may seem that it's going to be difficult. The Moon is a thousand times farther away than low Earth orbit, and Mars is two million times farther away. Luckily, we already do the most expensive part routinely. By far the biggest energy cost of going anywhere is prying ourselves away from the grip of Earth's gravity. The energy cost of getting to Earth orbit is 60 percent of the energy cost of getting to the Moon's surface and 50 percent of the energy cost of getting to the Mars surface. With its reusable orbital rockets, SpaceX already has the problem half solved!

The potential benefits of journeys to the Moon or settlements on the Moon are many and varied. They include scientific advances that come from zero- and low-gravity research environments, unrivaled opportunities for astronomers, a base from which to explore deeper into the solar system, and of course, incredible vacations. And yet it's over fifty years since a human set foot on the Moon. Apollo seems more like a dream than a majestic technological achievement. Anyone can be forgiven for thinking we've turned our backs on space. But the world's major space powers have detailed plans to establish a presence on the Moon.

In 2020, NASA launched the Artemis Accords, an international agreement to return humans to the Moon by 2024, set up a crewed lunar base by 2030, and cooperate on other space exploration goals.[12] NASA states its intention to land the first woman and person of color on the Moon. It awarded a $2.9 billion contract to SpaceX to develop a version of its Starship rocket to take US astronauts to the Moon for the first time since Apollo. The accords try to adhere to the 1967 UN Outer Space Treaty, which banned weapons in space and said activity in space should be for the benefit of all nations.[13] NASA also adheres to the United Nations' 1979 Moon Agreement, which prevents commercial exploitation of space resources—but this document wasn't signed by the world's major space powers.[14] Twelve countries have signed the accords, but they've been criticized as too centered on US and commercial interests.[15] NASA is collaborating with SpaceX on going to the Moon, and ESA has detailed plans for a lunar base.

Meanwhile, Russia and China are cooperating on space station and Moon missions.[16] Their plans reflect the geopolitical reality that they're both increasingly estranged from the West. Cooperation suits both sides—China gets access to highly experienced Russian space scientists, and the cash-starved Russian space program gets an injection of youthful energy and cash from China. A fierce new space race is brewing.

Figure 24.1
Plans for a lunar base involve bringing as little material as possible from Earth, due to the expense. The lunar soil can be used as protection from radiation, and 3D printing techniques will allow many construction components to be created from the soil. (Credit: ESA, "Lunar Base Made with 3D Printing," 2013, https://www.esa.int /ESA_Multimedia/Images/2013/01/Lunar_base_made_with_3D_printing.)

Getting to Mars is a much longer and more dangerous journey, even if the energy cost is similar. Astronauts may well experience a lifetime's worth of radiation exposure on the trip, and the level of isolation will be extreme. It will take seven or eight months to get there, as opposed to a few days to reach the Moon. If there's a medical emergency, astronauts will have to rely on the equipment they bring with them. After the first short-term visits, most experts think it will take a couple of decades to create enough infrastructure for a self-sustaining settlement.[17] The NASA plans call for the first landings by the end of the 2030s. Musk is famously optimistic about his time scales, but he insists that SpaceX will beat both China and the United States in putting people on Mars. He's also blunt about the risks: "Honestly, a bunch of people will probably die in the beginning," he said to X-Prize founder Peter Diamandis. "It's an arduous and dangerous journey where you may not come back alive, but it's a glorious adventure."[18]

Beneath Musk's bluster and hype is a steely resolve. Recently SpaceX has been hitting some impressive milestones. A longtime SpaceX employee said Musk's manic approach pays off: "It went from, *Holy crap, how are we going to do this?* to what I would consider a quiet professional confidence."[19] The key to his Martian dream is the Starship.

The SpaceX Starship is hugely ambitious—a game changer in the exploration of space. Originally code-named BFR for "big f*cking rocket," the Starship is the largest rocket ever built, eclipsing the Saturn V and 30 percent taller than the Statue of Liberty. The Starship is fully reusable. Its first stage is a booster powered by twenty-nine engines burning liquid methane and liquid oxygen. Its second stage is a long-duration spacecraft designed to take cargo and people to the Moon, Mars, and beyond.[20] The Starship can lift 110 tons to Earth orbit. Following a decade of development, the first test flights started in 2020, and the Starship nailed its first landing in 2021.

Figure 24.2
Artist's impression of the SpaceX Starship at separation from its heavy booster rocket. The Starship is a vehicle that will take astronauts to the Moon, in collaboration with NASA, and eventually to Mars. (Credit: SpaceX, 2018, https://commons.wikimedia .org/wiki/File:BFR_at_stage_separation_2-2018.jpg.)

After four flights that ended in fiery explosions, the fifth rocket successfully landed after reaching an altitude of ten kilometers (six miles). The marginal cost per flight is estimated to be five million dollars, lowering launch costs to twenty-five dollars per kilogram, one-thousandth the cost of cargo aboard the space shuttle. SpaceX will build dozens of Starships to take advantage of economies of scale and eventually launch them daily. That will require a great deal of methane and produce a great deal of carbon dioxide, while the rest of the planet is trying to reduce the use of fossil fuels and the concentration of carbon dioxide in the atmosphere. Concerns about energy use and climate change aside, daily flights by Starships could put 20 million tonnes in orbit every year. For reference, the International Space Station, which took a decade to build, has a mass of 420 tonnes. Starships could lift into orbit the equivalent of fifty thousand space stations per year! Not all the cargo will be cheese, caviar, and construction equipment. Even Musk bows to the rocket equation, so many Starship flights will be hauling methane and oxygen to refuel already-orbiting Starships for flights to more distant destinations. For landing on Mars, the Starship will use the atmosphere to decelerate, taking its steel heat shield close to its melting point and undoubtedly giving the passengers an extra thrill.

Casey Handmer, a prolific blogger about space travel, summarizes the importance of SpaceX's rocket: "Starship is a devastatingly powerful space access and logistical transport mechanism that will instantly crush the relevance of every other rocket ever built. Starship fundamentally changes our relationship with space."[21] Musk is building the capability to send a million people to Mars. Really.

What will they do there and how will they live? People will volunteer or pay for the hazardous journey to another world for the same reason they go to remote and wild places on Earth: to have a singular experience and push themselves to their limits.

Most of the technologies needed to live on the Moon or Mars already exist. Both the Moon and Mars have substantial ice deposits under the surface, so water can be extracted. Water can be separated into oxygen and hydrogen, and hydrogen is a key component of rocket fuel. Oxygen for breathing can be liberated from the regolith; lunar soil is nearly half oxygen by weight. Simple electrochemical processes can turn the soil into building materials. The Moon and Mars have little protection from high-energy radiation from space, so bubble domes would be covered with slump blocks made from the soil for protection from radiation and micrometeorites. It may be more efficient to build a habitat underground. Both the Moon and Mars have lava tubes tens of meters (tens of yards) deep and a hundred meters (three hundred feet) across with "skylights," where little excavation would be required to create a large living space.[22] ESA has been testing 3D printing using lunar soil to make building materials.[23] Three-dimensional printing has advanced so rapidly it can make many infrastructure components, and even replacement organs and body parts for astronauts.[24] Space settlers will undoubt-edly be technology innovators and early adopters, so they're likely to use genetic modification or biohacking to accelerate their adap-tation to an alien environment. Settlers will have to forgo their familiar comfort foods. The vision for feeding a million people on Mars will depend on eating insects and growing sources of nutrition in petri dishes.[25]

As for the relevance of these experiments for the rest of us stuck on Earth, the prime directive of space settlers will be the efficient use of resources. Water, air, clothes, and building materials—everything must be conserved and recycled. Social cohesion will be essential to survive such unforgiving environments. Although the experiment itself is grandiose, day-to-day life will be highly constrained. We can hope that lessons from the first space settlers will help us live more parsimoniously on Earth.

25

Mining Asteroids

Space offers many opportunities beyond those afforded by travel to and settlement on the Moon and Mars. One of the most obvious, with enormous potential rewards for anyone daring enough to exploit it, is the mining of asteroids.

"The first trillionaire there will ever be is the person who exploits the natural resources on asteroids." These are the confident words of astronomer Neil deGrasse Tyson.[1] The headline of an article on Bloomberg *is equally emphatic: "We're Never Going to Mine the Asteroid Belt."[2]*

Which will it be: feast or famine, rags or riches?

We're using the resources of the planet at an unsustainable rate. A graphic example of this is "Earth Overshoot Day," the day in the year when the human demand for ecological resources exceeds what the planet can regenerate in a year.[3] This is a rough, even controversial measure, but it's worked out by comparing the rate at which we consume resources with how fast the planet regenerates its resources. In 1987, Earth Overshoot Day was October 23; in 2021, it was July 29. Put another way, in 2021 we used 1.75 Earth's worth of biological resources. The depletion of hydrocarbon resources—coal, oil, and gas—is not necessarily bad if that helps accelerate a shift to renewable energy sources. In addition, crucial ingredients of the modern technological world—metals and rare earth minerals—are being used at a rapid rate.[4] When we use fossil fuels, we're drawing down the reserves of formerly living creatures deposited underground over hundreds of millions of years. With heavy elements, we're drawing down resources created in the cores of stars before Earth formed. Until we can master fusion, heavy elements are also nonrenewable.

It does not follow, however, that we're about to "fall off a cliff" in the accessibility of these vital ingredients of modern life. A Saudi oil minister once said, "The Stone Age did not end for lack of stone, and the Oil Age will end long before we run out of oil."[5] As the known reserves of any resource are depleted, the basic laws of supply and demand kick in. Prices will rise, and that can cause previously marginal or unknown reserves to be put into production. Substitution and recycling can ease the demand, and technology and lifestyle changes may mean that the future society's needs are different from today's. To take a few historical examples, flint was highly prized for arrowheads during the Stone Age, but the mining of flint has since stopped, and salt was valuable to the Romans, but its importance to the world economy has since declined. Predictions about the exhaustion or peak production of resources have usually been wrong.[6] The message here is that if you get a prospectus about asteroid mining in the mail and are thinking of investing, caution is advised.

To date, no asteroid has been mined. Yet two Japan Aerospace Exploration Agency missions have gathered small samples of asteroids and returned them to Earth. A third mission, NASA's OSIRIS-REx, has collected considerably more material than the other two—at least four hundred grams (fourteen ounces)—and is due to return it in 2023. The goal of that mission is scientific research, which is just as well since the net cost of $2.5 billion per kilogram wouldn't make for a good economic model.

Estimates of the value of asteroid resources generate eye-popping numbers. A carefully selected near-Earth asteroid five hundred meters across contains about $2 trillion worth of precious metals and $3 trillion worth of rare earth minerals.[7] Metals like gold and platinum are central to the electronics industry, and the seventeen rare earth elements near the bottom of the periodic table are essential in cell phones, car batteries, TV and computer displays, high-powered magnets, and a plethora of industrial and high-tech gadgets.[8] All of these heavy, iron-loving elements sunk to Earth's core during its molten youth over four billion years ago, leaving them rare in the crust. A subsequent hail of asteroids dusted the crust with heavy elements, and those are the deposits we mine today. If we bring asteroid material back to Earth, we'll be using technology to augment what nature provided eons ago.

Another type of asteroid contains large amounts of water. There are a thousand water-rich asteroids relatively near Earth, and each of the two dozen largest holds ten thousand Olympic swimming pools' worth of water.[9] Water does not have a high value on Earth, but it's essential in space to drink, hydrate food, and act as a radiation shield. Water can be split into hydrogen and oxygen, with both contributing to rocket fuel and to oxygen being available to breathe. Water is heavy and costs a good deal to lift into Earth orbit, so harvesting it in space is attractive.

Back in 2013, researcher Martin Elvis estimated the number of potential mining targets with the current technology. He accounted for the number within reach of today's rockets, the feasibility of mining them, and whether they would yield a profit. He calculated that ten metal-rich asteroids and eighteen water-rich asteroids are within our grasp.[10] He guessed that advances made by SpaceX over the past decade will increase those numbers by a factor of ten. The best nearby target is Anteros, a 2-kilometer-wide (1.2-mile-wide) asteroid named after the Greek god who avenged unrequited love. Anteros approaches within twenty-five times the Earth-Moon distance, and has an estimated metal and mineral value of a whopping $5.6 trillion.[11] But the most seductive target is 16 Psyche, often shortened to just Psyche. Named after the Greek goddess of the soul, Psyche is 220 kilometers (140 miles) across and orbits an average of 450 million kilometers (280 million miles) away in the asteroid belt.

Images and other data from ground- and space-based telescopes show
it's made almost purely of metals, with a market value of ten thousand quadrillion dollars, or a hundred thousand times the world's
economy.[12]

With stakes this high, the issue of ownership is important. Lassoing an asteroid and mining it for gold sounds like the Wild West.
Will there be any sheriffs out there?

The foundation of space law is a set of five treaties and five sets of
principles developed by the United Nations through its Committee
on the Peaceful Uses of Outer Space.[13] After ten years of negotiations,
the Outer Space Treaty became the first constitution for outer space in
1967.[14] It was ratified by ninety-nine countries and signed by an additional twenty-seven states. It says that space is the "province of mankind," and all nations have the freedom to "use" and "explore" outer
space, provided it's done in a way to "benefit all mankind." Some of
its sweeping and vague terms have never been clearly defined. Specific
issues were addressed by three subsequent treaties: a "Rescue Agreement" dealing with the return of astronauts from deep space, a "Lia-
bility Convention" to address any damage caused by space objects,
and a "Registration Agreement" to keep track of all objects launched
into space. The Moon Treaty, finalized in 1979, declares the Moon
to be part of the "common heritage of mankind" and says lunar
resources should be shared among all nations, and those resources are
"not subject to national appropriation by claim of sovereignty, by use
or occupation, or by any other means."[15] The treaty covers asteroids
as well as the Moon, but it's a "dead letter" since none of the major
space-faring powers have signed it.

All of these laws assume space is a realm dominated by countries,
not by companies, let alone by individual billionaires. As space
industrializes, it's obvious the law has some gaping holes; we've
seen one of them in the problem of space junk.

While international space law languishes, countries have been
acting to protect their interests and those of their citizens. In

November 2015, President Barack Obama signed the US Commercial Space Launch Competitiveness Act. The law recognizes the right of any person in the United States to own any space resources they harvest and encourages the commercial exploitation of asteroids.[16] In 2020, President Donald Trump went further, signing an executive order that said the United States doesn't view space as a "global commons." Luxembourg, a tiny European country with outsize ambitions, has passed legislation allowing firms incorporated there to carry out space mining.[17] NASA is trying to balance encouraging entrepreneurial activity with being a good global citizen. Its Artemis Accords (the international agreement to return people to the Moon by 2024) are seen by some as a power play by Western countries to control space commerce. Before the accords were signed, NASA administrator Jim Bridenstine tried to sound reassuring, affirming Article II of the Outer Space Treaty, which prohibits national sovereignty over objects in space: "We also believe that, just like in the ocean, you can extract resources from the ocean. But that doesn't mean you own the ocean. You should be able to extract resources from the Moon. Own the resources but not own the Moon."[18]

Not everyone's concerns are assuaged. The privatization of space could end up serving the interests of a small number of space capitalists rather than benefiting humankind. An extension of capitalism into space might repeat ills we face on Earth: the overexploitation of precious resources, power of monopolies, and corruption of crony capitalism. Listen to this critique by two social scientists, who contrast an activity they designate "NewSpace" with the "Old Space" mode of operations during the Cold War: "Despite its humanistic, universalizing pretensions, NewSpace does not benefit humankind as such but rather a specific set of wealthy entrepreneurs, many of them originating in Silicon Valley, who strategically deploy humanist tropes to engender enthusiasm for their activities."[19] These are real concerns, and if asteroid mining is to go ahead, much more work will be needed to avoid the self-serving excesses, human rights abuses, and environmental damage that capitalism can bring with it. Let's table this essential debate for now and see how asteroid mining would work.

Target selection comes first. Remember, all that glitters is not gold. While Psyche might contain vast potential wealth in precious metals, it's never less than about 250 million kilometers (155 million miles) away. Space miners start with the twenty-two thousand asteroids whose orbits bring them close to Earth. Then they investigate the subset that is "easily recoverable" because a relatively small change in their velocity could place them in orbit around Earth or the Earth-Moon system. Mining them close to Earth greatly reduces the cost of bringing extracted material home. In 2013, NASA proposed the Asteroid Redirect Mission, aiming to rendezvous with a large, near-Earth asteroid and use robotic grippers to retrieve an eight-meter boulder from the asteroid. The mission was also going to demonstrate a planetary defense technique by attaching thrusters to the large asteroid so that it could be steered away from a trajectory that might hit Earth. The same technology could be used to put the asteroid in a captive orbit.[20] The mission was canceled by the White House in 2017, however, leaving the private sector to pursue its dreams of mining without any government support. The private sector started modestly, identifying a dozen asteroids ranging in size from 2 to 20 meters (7 to 70 feet) that could be brought into an Earth-accessible orbit with an alteration in their velocity of under 500 meters per second, or 1,100 mph.[21]

The thought of space "cowboys" lassoing asteroids and deliberately bringing them near Earth might make you nervous. NASA rocket scientists, by contrast, are motivated by redirecting asteroids to protect Earth from potential impacts. In October 2021, NASA launched its Double Asteroid Redirection Test spacecraft to hit the smaller of two co-orbiting asteroids and change its orbit slightly—a proof of concept for missions that may one day be needed to save humanity. Orbital engineering shouldn't be undertaken lightly, and

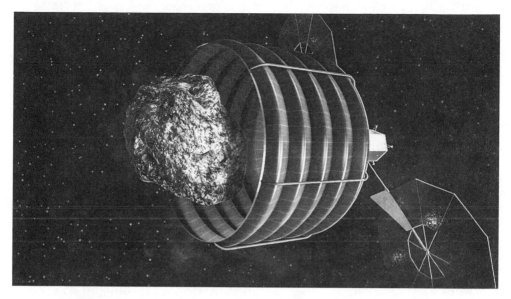

Figure 25.1
Asteroid mining will develop in stages. This NASA mission, not yet funded, was designed to capture a boulder up to eight meters across from a near-Earth asteroid and bring it into a captive orbit of Earth. (Credit: NASA, n.d., last edited 2015, https://commons.wikimedia.org/wiki/File:Asteroid_capture.jpg.)

since the fate of humanity could be in jeopardy, it should definitely be regulated![22]

An in-depth study of the feasibility of asteroid mining was commissioned by the Keck Institute for Space Studies at Caltech in 2013. The report saw no fundamental obstacles to the endeavor. It put a price tag of $2.6 billion on bringing a 7-meter, 550-ton asteroid into a high lunar orbit. That sounds like a lot, but the cost of starting a rare-earth-metal mine on Earth is around $1 billion.[23] The investment firm Goldman Sachs was bullish on asteroid mining in a report it issued in 2017.

Terrestrial mining methods won't work in the low-gravity environment of an asteroid. One innovative approach is called optical mining—using concentrated sunlight to drive water and other volatile materials from a water-rich asteroid while it's in an inflatable containment bag, avoiding the need for complex robotics. The TransAstra Corporation is designing a suite of spacecraft to locate and harvest asteroids: Mini Bee to demonstrate the light extraction method, Honey Bee to harvest asteroids up to 10 meters (35 feet) in diameter, and Queen Bee to harvest asteroids up to 40 meters (130 feet) in diameter.[24] Another is biomining. Biomining uses microbes to leach metals and rare earth elements from rocks. Microbes have mined Earth rocks to get nutrients for billions of years, and a lot of copper and gold is mined this way. Recent experiments on the International Space Station showed that biomining works just as well in microgravity as it does on Earth.[25]

Despite all the promises and hype, asteroid mining is off to a rocky start. Planetary Resources was founded in 2012. It boasted an A-list team that included film director James Cameron as an adviser, and Google CEO's Larry Page and Eric Schmidt as investors. The company had a long-term plan to establish a robotic asteroid-mining industry. It planned to create a fuel depot in space by 2020 using water extracted from asteroids. Instead, in 2020 the company was wound down and its assets were auctioned off. If you'd gone to the auction, you could have snapped up a thermal vacuum chamber for seven hundred dollars or satellite ground station for a thousand. There have been other casualties as well.[26] The nascent space-mining industry has a serious cash flow problem. There are riches in space, but deep-pocketed investors will be needed to get started.

It's a fact that terrestrial resources are finite and by comparison asteroid resources seem nearly unlimited. But the cost of setting up the infrastructure to mine and refine ore in space can only be speculated. The cost of returning the material to Earth might well exceed its market value. Even if it doesn't, the law of supply and demand will still operate. All the gold ever mined would fit into a container twenty meters on a side, equivalent to three Olympic swimming pools.[27] With gold hovering around two thousand dollars per ounce, it seems like there's plenty of money to be made.

Yet the demand for gold also matters, as does the marginal cost of extracting it on Earth. A company might dominate the market, but at the cost of cratering the value of the resources it had worked so hard to extract.

Entrepreneurs have a gleam in their eye, but all that glitters may not be worth its weight in gold. Fortunes will be made, but many will lose their shirts. This adventure will play out in the next thirty years.

26
Living beyond Earth

We shall not cease from exploration
And the end of all our exploring
Will be to arrive where we started
And know the place for the first time.
Through the unknown, remembered gate
When the last of earth left to discover
Is that which was the beginning.

—T. S. Eliot, "Little Gidding"

Eliot's quote is from an interleaved set of meditations on time, the universe, and the divine. It echoes the hero's journey, a pattern familiar in human history, mythology, and storytelling. The hero travels to unfamiliar lands and confronts good and evil in a personal journey where they discover themselves.[1] The journey can be thought of as the boundary between the two halves of the familiar yin-yang symbol of ancient Chinese folk philosophy. The symbol has a black swirl symbolizing yin and a white swirl symbolizing yang. These two can represent any opposing or complementary pairs, such as chaos and order. Each swirl contains a dot of the opposite color; one dot could signify hope amid mayhem, and the other reminding us that in calm without challenge there is no improvement. This interpretation of the symbol provides a useful reflection on exploration in general and space exploration in particular.

Astronomy teaches us that the universe is generally hostile to life, with isolated pockets of habitability. Persistence isn't guaranteed;

Earth has been uninhabitable in the past and it will be again in the future. If sentient creatures exist on far-flung exoplanets, they face the same issues we do. Our planet is under siege from its alpha species. Living off Earth may not be a panacea but it also doesn't have to be wasteful self-indulgence. The challenge will be to temper our restless urge to explore with wisdom.

Fewer than seven hundred people have ever been in Earth orbit, and only twelve have set foot on another world. Seven space tourists spent twenty to twenty-five million dollars each for a trip a decade ago. Space tourism surged in 2021 with a dozen civilians going up. For most of them, the cost was absorbed by the fledgling space companies. For the few who paid, the price tag was millions of dollars. Will space ever be an experience for other than a privileged few?

The trajectory of the commercial space industry suggests that space tourism will take off in the next few years and suborbital joy-rides will become common for wealthy people. We'll almost certainly build settlements on the Moon and Mars, and people will routinely live and work in space. The challenge is to avoid a vision of dystopic science fiction since *The Time Machine* by H. G. Wells where civilization is starkly divided into the haves and have-nots, a majority who live on the degraded Earth and a small minority who live in luxury off Earth.[2] This theme was also played out in the 2013 movie *Elysium*. Wells's novel was set in Victorian England, but his themes of inequality and class prejudice still pervade modern society. Exploration of space will only have value to our species if it plays a role in solving our chronic problems of sustainability and environmental degradation.

Traditional "hard" science fiction hews to narratives that mirror the development of technological societies. It's generally unquestioned that technology is a positive force. We see exploration and competition for resources play out in a grand arena off Earth. Social and cultural contexts rarely feature prominently. But there's a science fiction tradition that puts human concerns front and center.

It does not flinch from the dystopias that might await us, but finds optimism in a vision of cooperation and equality. Ursula K. Le Guin and Octavia E. Butler epitomize this strain of science fiction. In *The Dispossessed*, Le Guin explores a scenario where a population of humans has abandoned a capitalist planet for an anarchist society on a nearby moon.[3] In *Parable of the Sower*, Butler envisages a future of Earth devastated by climate change and racial violence where hope is found in embracing the inevitable chaos of change.[4] These authors and their young successors challenge us to build a noble future for humanity off Earth.

On Earth, sustainability must focus on the city. Cities occupy just 3 percent of the land area, but hold 55 percent of the world's population, generate 75 percent of the world's carbon emission, and use 80 percent of the world's energy.[5] Most big cities are sprawling, chaotic labyrinths, choked by traffic and shrouded in smog. There are attempts to retrofit cities with public transit and green technology, supported by the United Nations and World Bank, but it's an uphill battle.[6] The current exemplar is Copenhagen, which is on track to be the first carbon-neutral capital in 2025. All public transport is electric, and most trips are taken on bicycles, which greatly outnumber cars. From recycling to roof gardens to efficient heating and cooling, Copenhagen is addressing sustainability on all fronts. In the twentieth century, iconic architects Le Corbusier, Frank Lloyd Wright, and Buckminster Fuller designed domed cities, but none of their plans was built. In the Arabian desert outside Abu Dhabi, the city of Masdar is rising. Intended as a one-square-mile showcase of sustainability for fifty thousand people, it's far over its twenty billion dollar budget, many years behind schedule, and nearly as empty as a ghost town.[7]

These projects are far from being self-contained. A truly closed ecological system would be a habitat that didn't rely on any matter or energy exchange outside the system—something that the extreme parsimony required for habitation off Earth will necessarily teach us. Few experiments like this have been attempted. The Institute of Biophysics in Russia built an underground structure for three people in 1965 and ran the project for eight years. The Biosphere 2 experiment in southern Arizona ran for two years from 1991 to 1993. Eight scientists lived in the soaring glass structure, which contained five miniature habitats: a tropical rain forest, savanna grassland, mangrove wetland, coastal fog desert, and ocean with a coral reef. Biosphere 2 attracted a lot of publicity, not all of it favorable. The project was funded by (yet another) eccentric billionaire, Ed Bass. The team carried out unorthodox training with Aborigines in the Australian outback, and personal dramas colored the interactions of the occupants. Late in the experiment, carbon dioxide levels fluctuated wildly, some insect populations crashed while others turned into plagues, and the scientists were forced to start eating their emergency supplies. It's suspected that one snuck out and smuggled in energy bars.[8]

For unforgiving environments like the Moon and Mars, habitats must be sealed and the cost of delivering supplies is prohibitive, so recycling and self-sufficiency are essential. NASA published a book on urban planning in space in 1977.[9] Currently, a team of architects is working with ESA on a "Moon Village" concept. The village would be an expandable settlement at the rim of the Shackleton Crater near the South Pole, a place with water ice and nearly continuous sunlight. Inflatable domes would be covered by protective shells, constructed by robots, and 3D printed using lunar soil. An organization called Mars City Design, founded in 2015, hosts competitions that aim to stimulate creative thinking about sustainable living on Mars.[10] In 2017, an MIT team won the competition. Domes and tree habitats, each housing fifty people, would be connected by a series of tunnels. The team leader elaborated on the forest metaphor: "On Mars, our city will physically and functionally mimic a forest, using local Martian resources such as ice and water, regolith (or soil), and Sun to support life. Designing a forest also symbolizes the potential for outward growth as nature spreads across the Martian landscape. Each tree habitat incorporates a branching structural system and an inflated membrane enclosure, anchored by tunneling roots."[11]

Truly closed ecological systems are the key to living off Earth, but they also have enormous potential applications here. If pollution and the destruction of fragile ecosystems continue unchecked, these habitats will be crucial for sustaining life on Earth.[12] In the science fiction novel *2312*, Kim Stanley Robinson envisions a future where we've inhabited other planets in the solar system. We turn our attention to the ravaged Earth, and restore and "rewild" the planet by returning species that have been preserved in off-Earth sanctuaries.[13] Robinson has had ecological themes for most of his novels. He is unique among his peers in the way he envisions moral not just technological progress. His protagonists fight to keep future societies from repeating our mistakes.[14]

If we move off Earth in large numbers, ethical issues will need to be addressed. Propagating life beyond our planet raises questions: "Should we expand all life or only intelligent life? Which humans will get to leave Earth? Who makes that decision? Will we allow the population beyond Earth to grow without limit? Are the space settlers subject to terrestrial laws or will they make their own laws? Should we alter biology off-Earth to suit our needs, or only seed life on dead planets, or not alter nature off-Earth at all? Should we seed life in other planetary systems? How far can we change our biological makeup while still preserving the human species, and life itself?"[15]

Imagine enormous spheres orbiting Earth, spinning to create artificial gravity, with tens of thousands of people living on a curved inside surface covered with houses, parks, and rivers, like an "inside-out planet." Welcome to a habitat called Island One, the vision of Gerard K. O'Neil.

O'Neil was trained in physics and spent most of his career at Princeton University. He worked in high-energy physics and made a major impact with his invention of the particle storage ring. Yet

Figure 26.1
Cutaway view of a space habitat, in an artist's impression, based on designs by Gerard K. O'Neil. The giant wheel is in Earth orbit and spins slowly to create artificial gravity. (Credit: NASA/R. Guidice, 1970s, https://commons.wikimedia.org/wiki/File:Stanford_Torus_cutaway.jpg.)

he soon became consumed with the idea of how to live in space, exploring the notion through a seminar for his students in 1969. Like many space visionaries, he took his cue from a Russian we've already encountered: "The Russian pioneer of space travel, Konstantin Tsiolkovsky, came closest to suggesting an earthlike environment with his greenhouses. His excellent designs, put forth seventy-five years ago, were basically tubular and very efficient. He had a lot of the essential ideas right: to go for unlimited, clean solar energy outside the planet's shadow; to make use of the resources from asteroids. Aside from him, almost everyone thought of space as a route from here to there. The destination was always assumed to be a planetary surface. But once you say that space itself can be the destination rather than just a corridor—that you can build large, earthlike environments in space—you get a radical change in viewpoint."[16]

While these structures might seem confining, O'Neil saw them as liberating and expanding our freedoms: "First of all, there would be fewer people living on Earth and an increasing fraction living in space, where there's unimaginable room. Those in space settlements would of course find the situation much more open and freer. They'd be living in relatively small-scale structures, in habitats that would be community-size rather than nation-size. With a few thousand to perhaps fifty thousand people in each space settlement, government could be as simple and intimate as a New England town meeting. Yet each settlement could be quite self-sufficient, using pure solar energy to generate power for travel, agriculture, environmental control, and so on. Since the settlement would be growing its own food, there would be no reason for it to tie into a large-scale governmental structure."[17]

O'Neil contended that building space settlements would solve several critical problems: "It is important to realize the enormous power of the space-colonization technique. If we begin to use it soon enough, and if we employ it wisely, at least five of the most serious problems now facing the world can be solved without recourse to repression: bringing every human being up to a living standard now enjoyed only by the most fortunate; protecting the biosphere from damage caused by transportation and industrial pollution; finding high quality living space for a world population that is doubling every 35 years; finding clean, practical energy sources; preventing overload of Earth's heat balance."[18]

Bezos was inspired by O'Neil's designs. At a media event in 2019, the Amazon founder and owner of Blue Origin described his vision of a trillion people living in space: "This would be an incredible civilization," he said. "We will run out of energy. This is just arithmetic. It's going to happen. Do we want stasis and rationing, or do we want dynamism and growth?"[19] Bezos can display hubris and arrogance, but his point about the arc of our energy usage is valid. Orbiting settlements can be combined with asteroid mining. By building the space settlement around an asteroid that was already being mined, the asteroid can provide water, plus aluminum, titanium, and other construction materials. Legions of robots will do the work. These habitats are to the International Space Station as Yellowstone is to a city park.

Entrepreneurs with the gumption to start a space business will be willing to invent new technologies, so we can expect to see a lot of innovation off Earth. They'll extract water from rocks for drinking, growing crops, and acting as a radiation shield, and to provide hydrogen and oxygen, the fuel and oxidizer for rocket engines. They'll use 3D printing to fabricate machines from scratch from local materials.

Expect to see space as the place where genetic engineering is pushed, including the settlers who hack their genomes to improve their adaptation to alien environments. They'll use nanobots injected into their bloodstream to monitor their health and make continuous adjustments as well as do minor internal repairs. Space settlers will probably achieve the goal of a John von Neumann machine, an autonomous robot that can reproduce itself using materials found in the space environment. This capability would allow humans to rapidly expand their capabilities off Earth.

The most grandiose vision for living off Earth is to make another planet habitable. If we could populate all of Mars, rather than just living in a few pressurized bubble domes, it would be the ultimate hedge against our extinction on Earth. This sounds like the ultimate hubris, to bend an entire planet to our will. Is it actually possible?

Mars is at the outer edge of the habitable zone, with a thin atmosphere of carbon dioxide and a surface of cold, arid deserts. Three billion years ago, Mars was warmer, had a thicker atmosphere, and was covered in lakes and shallow seas. To restore the red planet to its former glory would be a herculean feat of geoengineering, costing trillions of dollars and probably taking hundreds or thousands of years. The goal is to build up the atmosphere and heat the planet. If the surface temperature could be raised by 20–30°F, liquid water will persist on the surface. There's a huge inventory of frozen carbon dioxide near the poles; once it's liberated, the greenhouse effect will kick in, accelerating the process.

The practicalities of terraforming are intimidating but not impossible. Ideas to raise the temperature include deploying orbital mirrors to deliver additional sunlight and the efficient but risky method of detonating nuclear weapons on the surface. As mentioned earlier, Musk has been vocal about this method as the way to achieve his goal of populating Mars, earning him a reputation as a kind of James Bond villain.[20] But even if we could liberate all the carbon dioxide in the Martian soil, it would only yield an atmospheric pressure of 15 percent of Earth's and only raise the temperature 10°C, not enough to sustain liquid water. That means importing more carbon dioxide by redirecting comets and asteroids to hit Mars, or using more efficient greenhouse gases like chlorofluorocarbons. Another challenge is making the atmosphere breathable. The MOXIE experiment on the Mars Perseverance rover has been demonstrating how oxygen can be generated from the soil, but the more efficient way is to genetically engineer organisms like Earth's cyanobacteria for that purpose.[21]

Even if we could create earthlike conditions, they won't last. Mars has weak gravity and no magnetic field, so the solar wind will steadily strip away its atmosphere and water. Amazingly, there may even be a fix for that. NASA chief scientist Jim Green has shown how an artificially induced magnetic field of ten thousand gauss, placed between the Sun and Mars at a place where their gravities cancel out, would protect Mars from the solar wind.[22]

Engineering the planet is one strategy for living off Earth. Engineering the people is another. In the movie *Gattaca*, only superhumans with altered genomes are allowed to travel to Titan, while the genetic losers, called "in-valids," can only watch as the rockets launch. Like all good science fiction, this 1997 film is not far from reality. CRISPR genome editing technology has put customized genetic modification within reach.[23] The genetic "fitness" of an organism is how well it thrives and reproduces in a particular environment. The fitness of a human off Earth is low. Veritas Genetics will sequence your entire genome and give you a report on your "space genes."[24] Juicing up your genome for living in space may include adding copies of p53, a gene involved in preventing cancer. Or adding the variant of EPAS1 that lets Indigenous Tibetans use less oxygen. Or including natural mutations associated with problem-solving skills and low anxiety, or with large, extra-lean muscles.[25] Presumably all space settlers will have or need the variant of the DRD4 gene associated with epic migration in early nomadic cultures as well as the risk-taking behavior in the current population.[26] DRD4 has been dubbed the "explorer gene." It may be part of the reason we left Africa fifty thousand years ago, and why we alone, among all species, restlessly explore our planet without being driven by the need for food or shelter.

Why stop there? Check the label of a box of cereal or frozen dinner and you'll see a list of "essential" nutrients and vitamins that the body can't make. That's why we must eat plants, fungi, and bacteria that can make them, called prototrophs. Prototrophs synthesize all they need from simple ingredients in the soil. About 250 new genes would have to be created to engineer a prototrophic human cell—a tall order.

Geneticist Christopher Mason is taking the long view. He lays out a five-hundred-year plan for engineering humans for space, dramatically accelerating the process of evolution: "The DNA in our cells works regardless of where it came from, meaning our gene networks do not choose their place in our cells with any regard to their history; rather, their place is determined based on what is needed. This same principle that applies to life on Earth can be readily applied beyond Earth as well. Given these widespread and pervasive examples of exchange of DNA between species, it is not unexpected, or unnatural, to begin to think about doing so in human cells. Because our own human lineage only provides evolutionary lessons from the past few million years, we would be better served by taking the lessons from billions of years of evolution for us to survive on faraway worlds."[27] The redesign might get radical. Mason has even speculated about engineering humans so we can photosynthesize, living off light instead of eating food. That means getting chloroplasts to work inside human skin cells. To produce enough energy to live this way, a person would need to be as thin as a leaf and the size of a playground. After such radical genetic engineering, a person would hardly be human anymore. Perhaps it's good that this brave new world is far away.

All of this effort is just to live elsewhere in our solar system. Beyond our harbor is the great blue sea. The nearest earthlike planets are millions of times farther away than Mars. We've seen that it's ambitious just to send nanobots to the nearest star. Living beyond our solar system won't happen in our lifetimes or our great-grandchildren's lifetimes. To reach the stars, we must transcend the limitations of chemical rockets. Some technologies could get us there, but none has been tested yet.[28] We'll never travel to the stars because we need a new place to live; it's far easier to create new habitats on or near Earth. If we go, it will be because of the restlessness and hunger that are part of the human condition.

Eventually, as novel propulsion methods are perfected a century or more from now, let's speculate that one of O'Neil's Island One spheres sets off for another part of the galaxy. Generations of inhabitants live their lives in transit but eventually reach another star.

In this way, we finally travel to some of the earthlike planets that astronomers have discovered. It will be the beginning for a new branch of humanity.

French aviator Antoine de Saint-Exupéry captured the poetry at the heart of the dream of space exploration: "One will weave the canvas; another will fell a tree by the light of his ax. Yet another will forge nails, and there will be others who observe the stars to learn how to navigate. And yet all will be as one. Building a boat isn't about weaving canvas, forging nails, or reading the sky. It's about giving a shared taste for the sea, by the light of which you will see nothing contradictory but rather a community of love."[29] The lesson: to motivate people to build a ship for a long voyage, teach them to yearn for the vast and endless sea. But to sail that sea, build a community that excludes no one. Only then will our love of the sea, and the cosmos, be fully realized.

Epilogue: Scenes from the Future

Expanding our footprint into space in the ways described in the last section of this book could take decades and trillions of dollars. And it might exacerbate the problems we already have by widening the gap between the super-wealthy and privileged and the extremely poor and disadvantaged, utilizing fossil fuels and other resources that we desperately need to keep in the ground, or bringing danger and poor conditions to workers to realize the dreams of the rich and powerful.

There's a strong argument for imposing a hiatus on human space exploration and focusing our efforts on the real challenges we face. We need to build a fairer world in which there's genuine opportunity, respect, and education for all, and find ways to live sustainably.

As I've already discussed in this book, however, there are many potential long-term advantages to stepping up space exploration, even if there's pain in the short term. And then of course there's that human instinct for exploration and our natural curiosity, both of which are unstoppable.

And so to close this book's narrative, I'll extrapolate current technology to a future where we've embraced the challenge of space and used it to live better lives on Earth. In this way, we can envisage possible futures for humanity. Rather than accept the limits of Earth, our exploration of space will let us expand the footprint of human civilization and create new habitable locations in our solar system, and eventually far beyond.

In ten years, space will no longer be the exclusive preserve of nations. As with the internet, the technology was incubated by governments and then the private sector took over. For the first time since the space age began with Sputnik, more space launches are being carried out by private companies than by governments. At the current pace, around the year 2030, the thousandth space tourist will ride a trembling rocket to the blackness of near-Earth orbit, and a hundred people will have spent a few nights in the Marriott Inflatable Space Module. Such extravagance is hard to defend in a world where seven hundred million live in extreme poverty and two billion do not have clean drinking water. In this rarified setting, we can imagine a wealthy couple booking a year in that unique hotel with their fees paid in exchange for streaming rights by Netflix or its successor. The moments of their daily lives will light up the internet. Perhaps they'll announce a pregnancy and near the end of their year in zero gravity, with the nearest hospital over two hundred miles away straight down, a brand-new life will begin, the first-ever off world.

In twenty years, we'll likely have a modest footprint on the Moon and Mars. On the Moon, an international consortium will build the infrastructure required to enable routine travel in our solar system. After the initial novelty, the public will lose interest in the daily routine of life on the Moon, but it'll be intrigued by the lunar elevator under construction. As a capital project, such an elevator will be no more costly than a large bridge or dam on Earth, and the first-mover advantage to companies wanting to be in the first wave to exploit lunar resources is irresistible. Like a rope trick, the slender structure will reach into the sky with no visible means of support. On Mars, NASA and SpaceX will operate side by side with mutual support necessary, but also an intense rivalry on display. Designs for the habitats are already in place. They'll be domes covered in bricks, like red-brown igloos perched on top of a steep escarpment with views of the endless dunes beyond.

With less certainty, we project another decade, to thirty years from now. The world will return, in fits and starts, to "bending the curve" of global warming. We're on track for half the world's energy coming from renewable sources. But while the industrialized world is mostly green, the less industrialized countries are still using fossil fuels, with growing demand as they transform their economies and their populations continue to move into megacities. Scientists already have radical plans to cool the planet by deflecting sunlight with a fleet of thin mirrors in space. Such technological solutions are a sign that humans are still prone to consumption not conservation. Geneticists are already racing to scan the genomes of rapidly disappearing species. These efforts will intensify. Technocrat billionaires like Musk, Bezos, and Branson conceive of visionary space projects. But a new, young technology guru might put her wealth into preserving the remaining biodiversity of flora and fauna. Svalbard has its seed vault, dug into Arctic ice, but this will be more ambitious: a Noah's Ark in Earth orbit, spinning to create artificial gravity, based on designs that Gerard O'Neill drew up for NASA in the 1970s. If our planet's ecosystem crashes, the ark will "rewild" Earth and spark a new beginning.

Forty years from now, if biology follows its current trajectory, genetic information will essentially be free. The power of genetic engineering will be in the hands of individuals. Progress will be made in fighting a swath of inheritable diseases, and "designer babies" will go from fantasy to reality. Settlers on the Moon and Mars will be selected to be technology savvy, so they aggressively adopt genetic modification aided by a lax regulatory environment off Earth. They will accelerate their rate of adaptation to environments with low gravity and no natural pathogens. But they, like all of us on Earth as well, will remain trapped in our solar system since the energy requirements to reach even the nearest star means that the journey will take centuries or millennia.

Meanwhile, space ventures will grow out of their buccaneering phase. Tourism in low Earth orbit will be within the reach of middle-class families. Space travel will follow the arc of commercial aviation, which was exclusively for the affluent in the 1930s, but within half a century reached the masses. The increase in activity in space will spur the United Nations to craft a new Space Treaty, the first since 1965, to affirm the ban on weapons in space as well as resolve issues of ownership and governance. Around this time, an orbiting solar collector will go online, ready to beam a gigawatt of microwave power to the surface. It's a prototype for arrays a hundred times larger that can put a big dent in our ravenous demands for energy. Its big advantage: in space, the Sun always shines. After a shakedown of the weaker players, a handful of asteroid-mining companies are the new unicorns of the financial sector, with valuations over a hundred billion dollars based on the promise of cheap minerals and precious metals.

Another prediction, for fifty years from now, is that Breakthrough Starshot works through its technical challenges, and aided by a billionaire's deep pockets and crowdsourced funding, it launches a fleet of miniature spacecraft to the nearest star. These small spacecraft are accelerated by solar sails gathering the Sun's rays and travel on fleet feet through the vacuum of interstellar space. Around 2075, with a few losses along the way, they reach their destination and transmit data back to Earth. After a four-year tantalizing wait, the data will reach us. The remote sensing will confirm astronomers' predictions: Proxima Centauri b is a virtual Earth clone in the star system next door. Gravity and air pressure are a little less than on Earth, and there's more oxygen and methane along with a bit less nitrogen and carbon dioxide. Spectra show a chlorophyll edge and water absorption, and the reflectance suggests a planet one-third covered with water. The evidence points to similar microbial metabolisms to those on Earth, but with no signs of technology or artificial structures. Almost immediately, space visionaries plan a voyage to this new world.

This leads to a last prospect, pure speculation yet based on an extrapolation of current technologies. A century from now, we imagine the world has been following the progress of *Prometheus* as it sails the dark, silent seas of interstellar space. The starship finally reaches Proxima Centauri b and a mechanical army springs into action. Construction robots replicate, using local materials to create a sealed habitat. Frozen embryos are taken to the surface and carried to term in artificial wombs. Cyborg nurses and nannies raise the pioneer children. As the cohort grows, the cyborgs convert into teachers to convey language, ethics, and cultural values. The home planet watches this bizarre reality show, with a four-year delay due to the finite speed of light. Debates rage. Is this a monstrous experiment or a tabula rasa, the fresh start for humanity? Either way, the step is just as momentous as when humans first ventured out of Africa fifty thousand years earlier. This bleeding-edge project has less well-publicized counterparts in Earth orbit and on the Moon, where new models of egalitarian societies are being explored. Populating space is an activity that pushes us outward while inducing introspection and motivating us to grow as a species.

Acknowledgments

It has been a wild ride to try to master the rapidly evolving field of exoplanet research. As someone whose research focuses on distant realms of the universe, I've relied on the insights and knowledge of numerous colleagues working in planetary science and astrobiology. These colleagues were always generous with their time and patient with my questions. Any errors that remain in the book are purely mine. I acknowledge the stimulating and productive environment of the Aspen Center for Physics, where major sections of this book were written. Also, a final coat of polish was applied at the Bellagio Center of the Rockefeller Foundation, during a residency that was optimally designed for thinking, writing, and sharing ideas. I'm grateful to my agent, Anna Ghosh, for advising me, and calmly guiding me through the opaque and sometimes choppy waters of the publishing industry. I've enjoyed working on this project with Jermey Matthews, my editor at the MIT Press. Finally, I'm grateful to my wife, Dinah, for her unstinting and enthusiastic support of my writing.

Notes

Prologue

1. Leibniz made the argument in three stages. First, God can only choose one universe from the infinite number of possible universes. Second, God is perfect and makes decisions based on reason. Third, the existing world, since it is chosen by God, must be the best. With regard to evil, God created the best possible world with the most good and least evil, commensurate with the existence of free will. Leibniz coined the phrase "best of all possible worlds" in a book of essays published in 1710.

2. "Panglossian" is used as an adjective to describe someone who is naively or unreasonably optimistic. Several other philosophers, including Immanuel Kant and Bertrand Russell, have critiqued the logic of Leibniz's argument.

3. "Darwin's Warm Little Pond Revisited: From Molecules to the Origin of Life" by H. Follmann and C. Brownson, 2009, *Naturwissenschaften*, volume 96, number 11, 1247–1263.

Part I

1. *Pale Blue Dot: A Vision of the Human Future in Space* by C. Sagan, 1997, New York: Ballentine Books, 6.

Chapter 1

1. Almost everything we know about Lucian comes from his own writings. Since his writing could be so hyperbolic, it is probably wise to take his accounts of his own history with a grain of salt.

2. *Lucian: A True Story, Collected Ancient Greek Novels* by B. P. Reardon, 1989, Berkeley: University of California Press, 619–649; "Lucian's True History as SF" by S. C. Fredericks, 1976, *Science Fiction Studies*, volume 3, number 1, 49–60.

3. *New Maps of Hell: A Survey of Science Fiction* by K. Amis, 1960, New York: Harcourt Brace, 28.

4. "Review: The Letter to Herodotus" by A. A. Long, 1974, *The Classical Review*, volume 24, 46–48.

5. "On the Heavens" by Aristotle, 350 BCE, http://classics.mit.edu/Aristotle/heavens.1.i.html.

6. Aristotle's cosmology became part of the dogma of Christian thought, but other religions were more open to the idea of many worlds. Medieval Muslim scholar Muhammad al-Baqir wrote, "I swear by God that God created thousands and thousands of worlds and thousands and thousands of humankind." Several verses in the Quran support the idea of multiple worlds, and they feature in one of the fictional stories from *One Thousand and One Nights*. Buddhism cosmology also involves multiple worlds and planes of existence. Many Buddhist traditions consider it uncontroversial and even inevitable that humans are not the only sentient creatures in the universe.

7. *Humble before the Void: A Western Astronomer, His Journey East, and a Remarkable Encounter between Western Science and Tibetan Buddhism* by C. Impey, 2015, West Conshohocken, PA: Templeton Press. See also "Cosmology: Myth or Science" by H. Alfven, 1984, *Journal of Astrophysics and Astronomy*, volume 5, 79–98.

8. "The British Industrial Revolution and the Ideological Revolution: Science, Neoliberalism, and History" by W. J. Ashworth, 2014, *History of Science*, volume 52, 178–199.

9. *Giordano Bruno: An Introduction* by P. R. Blum, 2012, Amsterdam: Rodopi.

10. "Giordano Bruno and the Heresy of Many Worlds" by A. A. Martinez, 2018, *Annals of Science*, volume 73, number 4, 345–374.

11. Quoted in "In Retrospect: Kepler's Astronomia Nova" by J. J. Lissauer, 2009, *Nature*, volume 462, 725.

12. *The Sidereal Messenger* translated by A. Van Helden, 1980, Chicago: University of Chicago Press.

13. "Kepler's Somnium: Science Fiction and the Renaissance Scientist" by G. E. Christianson, 1976, *Science Fiction Studies*, volume 3, number 1, 76–90.

14. *The World of Science Fiction, 1926–1976: The History of a Subculture* by L. del Rey, 1980, Shrewsbury, MA: Garland Publishing, 12.

15. *A Brief Literary History of Science Fiction* by R. Scholes and E. S. Rabkin, 1977, London: Oxford University Press.

16. "Science Fiction: A Brief History and Review of Criticism" by M. B. Tymn, 1985, *American Studies International*, volume 23, number 1, 41–66.

17. For a conventional view of the history of science fiction, see *The Cambridge Companion to Science Fiction* edited by E. James and F. Mendlesohn, 2003, Cambridge: Cambridge University Press. For an iconoclastic view, see *Trillion Year Spree* by B. Aldiss and D. Hargrove, 1986, New York: Atheneum Books.

18. "Futures Dreaming outside and on the Margins of the Western World" by I. Milojevic and S. Inayatullah, 2003, *Futures*, volume 35, 493–507.

19. "Math and Magic: Nnedi Okorafor's *Binti* Trilogy and Its Challenge to the Dominance of Western Science in Science Fiction" by B. Burger, 2020, *Critical Studies in Media Communication*, volume 37, 364–377. See also *Speculative Blackness: The Future of Race in Science Fiction* by A. M. Carrington, 2016, Minneapolis: University of Minnesota Press.

20. "Cosmic Concordance" by J. P. Ostriker and P. J. Steinhardt, 1995, arXiv:astro -ph/9505066.

21. "The Dark Side of Cosmology: Dark Matter and Dark Energy" by D. N. Spergel, 2015, *Science*, volume 347, 1100–1102.

22. "How Many Stars Are in the Universe?" by E. Howell and A. Harvey, February 11, 2022, Space.com, https://www.space.com/26078-how-many-stars-are-there.html.

23. The five planets visible to the naked eye have been known since antiquity: Mercury, Venus, Mars, Jupiter, and Saturn. Uranus was discovered by William Herschel in 1781, and Neptune was discovered by Urbain le Verrier in 1846. Pluto enjoyed decades as a planet after its discovery by Clyde Tombaugh in 1930, before it was demoted to the status of a mere dwarf planet by the International Astronomical Union in 2006.

Chapter 2

1. "A Jupiter-Mass Companion to a Solar-Type Star" by M. Mayor and D. Queloz, 1995, *Nature*, volume 378, 355–359.

2. *Introduction to Astronomical Spectroscopy* by I. Appenzeller, 2012, Cambridge: Cambridge University Press.

3. "A Planet Orbiting the Neutron Star PSR 1829–10" by M. Bailes, A G. Lyne, and S.L. Shemar, 1991, *Nature*, volume 352, 311–313; "No Planet Orbiting PSR 1829–10" by A. G. Lyne and M. Bailes, 1992, *Nature*, volume 355, 213. Ironically, the following year, two pulsar planets were detected and later confirmed, so they preceded the 51 Pegasi detection as the first discovery of planets outside the solar system.

4. "A Brief Personal History of Exoplanets" by P. Butler, 2016, *Pale Red Dot*, https://palereddot.org/a-brief-personal-history-of-exoplanets-by-paul-butler/.

5. "Interview: Michel Mayor, Nobel Prize-Winning Astrophysicist" by H. Lamb, 2020, *Engineering and Technology*, https://eandt.theiet.org/content/articles/2020/08/interview-michel-mayor-nobel-prize-winning-astrophysicist/.

6. "The Remarkable Race to Find the First Exoplanet, and the Nobel Prize It Produced" by M. Kaufman, 2019, Many Worlds, https://manyworlds.space/2019/10/10/exoplanet-glory-days-the-extraordinary-scientific-times-that-led-to-a-nobel-prize/.

7. "A Planetary Companion to 70 Virginis" by G. W. Marcy and R. W. Butler, 1996, *Astrophysical Journal Letters*, volume 464, 147–151.

8. "A Brief Personal History of Exoplanets" by P. Butler, 2016, *Pale Red Dot*, https://palereddot.org/a-brief-personal-history-of-exoplanets-by-paul-butler/.

9. "Disk-Satellite Interactions" by P. Goldreich and S. Tremaine, 1980, *Astrophysical Journal*, volume 241, 425–441.

10. "Origins of Hot Jupiters" by R. I. Dawson and J. A. Johnson, 2018, *Annual Reviews of Astronomy and Astrophysics*, volume 56, 175–221.

11. "Three Planets for Upsilon Andromedae" by J. Lissauer, 1999, *Nature*, volume 398, 659–660.

12. A more fundamental problem with all Doppler detection of exoplanets is the fact that only the radial component of the velocity is measured. If the orbit is in a plane parallel to the line of sight, the data gives the full velocity and mass, but if the orbit is in a plane perpendicular to the line of sight, there is no signal. For random orientations of the exoplanet orbit, the technique gives a minimum mass, and averaging over a large sample, planet masses are underestimated by a factor of two.

13. "Naming of Exoplanets," International Astronomical Union, https://www.iau.org/public/themes/naming_exoplanets/.

14. "Debra Fischer: All about Exoplanets," Superstars of Astronomy interview with Dave Eicher, 2015, astronomy.com, https://astronomy.com/-/media/Files/PDF/Superstars/Fischer.pdf.

15. "Nobel Lecture: Plurality of Worlds in the Cosmos: A Dream of Antiquity, a Modern Reality of Astrophysics" by M. Mayor, 2020, *Reviews of Modern Physics*, volume 92, 1–12.

16. The first Nobel Prize in Physics was awarded in 1901. Astronomy or astrophysics was not recognized until 1974, when Martin Ryle and Antony Hewish won for the discovery of pulsars. In the next forty years, there were six prizes for astronomy: Arno Penzias and Robert Wilson for the microwave background in 1978, William Fowler and Subrahmanyan Chandrasekhar for nuclear astrophysics in 1983, Russell Hulse and Joseph Taylor Jr. for binary pulsars in 1993, Riccardo Giacconi for X-ray astronomy in 2002, John Mather and George Smoot for the microwave background

in 2006, and Saul Perlmutter, Brian Schmidt, and Adam Riess for the accelerating expansion of the universe in 2011. Astronomy has netted the prize in three of the last four years. Apart from Mayor and Queloz in 2019, Rainer Weiss, Barry Barish, and Kip Thorne won for gravitational waves in 2017, and Andrea Ghez, Reinhard Genzel, and Roger Penrose won for massive black holes in 2020.

17. "A Planetary System around the Millisecond Pulsar PSR 1257+12" by A. Wolszczan and D. A. Frail, 1992, *Nature*, volume 355, 145–147.

18. The following years were bittersweet for Marcy and Butler. After fifteen years, their collaboration ended in acrimony. Butler continued as a scientist for life at the Carnegie Institution in Washington, but Marcy's fall from grace was swift and dramatic. In 2015, he was found guilty of multiple violations of the University of California's sexual harassment policy for over a decade, and a few months later he resigned from his faculty appointment.

19. "Interview: Michel Mayor, Nobel Prize-Winning Astrophysicist" by H. Lamb, 2020, *Engineering and Technology*, https://eandt.theiet.org/content/articles/2020/08/interview-michel-mayor-nobel-prize-winning-astrophysicist/.

Chapter 3

1. "William Borucki on the Power of Resolve in NASA's Kepler Mission" by I. Evans, 2019, *Elsevier Connect*, https://www.elsevier.com/connect/everyone-agrees-your-method-is-not-going-to-work-william-borucki-on-the-power-of-resolve-in-nasas-kepler-mission.

2. "Kepler Planet-Detection Mission: Introduction and First Results" by W. J. Borucki et al., 2010, *Science*, volume 327, 977–980.

3. The depth of the eclipse, or amount of dimming, is given by the ratio of the cross-sectional area of the planet to the area of the star. Jupiter is ten times smaller than the Sun, so taking the square, the ratio is 1/100 or 1 percent. Since some light will filter through the gaseous atmosphere of Jupiter, in practice the dimming is slightly less. Transits of the Sun by Venus were used to triangulate the distance to Venus and so estimate the distances to all the planets in the solar system during the seventeenth century.

4. The argument is based on simple geometry. The depth of the eclipse during a transit does not depend on how far away the planet is from its star. But the probability of seeing a transit for randomly oriented systems does. If the planet orbits close to the star, a transit will be seen for a wider range of orientation angles than if the planet orbits far from the star.

5. "Detection of Planetary Transits across a Sun-Like Star" by D. Charbonneau, T. M. Brown, D. W. Latham, and M. Mayor, 2000, *Astrophysical Journal*, volume 529,

45–48. See also "A Transiting '51 Peg-Like' Planet" by G. W. Henry, G. W. Marcy, R. P. Butler, and S. S. Vogt, 2000, *Astrophysical Journal Letters*, volume 529, 41–44. The two groups submitted their papers within a day of each other, and the papers were published back-to-back in the main astronomy research journal two months later.

6. "David Charbonneau Interview," 2017, *Harvard Crimson*, https://www.thecrimson .com/article/2017/9/14/fifteen-professors-2017-david-charbonneau/.

7. In microlensing, the foreground star is at a different distance than the background one, so its momentary passage in front of the more distant star causes a temporary brightening due to the general relativistic effect of mass bending light, and the signal never repeats. This distinguishes lensing from a binary star or a star plus planet where the brightness change repeats with each orbit.

8. "Farthest Known Planet Opens the Door for Finding New Earths," *Space Daily*, 2003, https://www.spacedaily.com/reports/Farthest_Known_Planet_Opens_the_Door_For_ Finding_New_Earths.html.

9. "A Brief History of the Kepler Mission" by W. J. Borucki, 2010, NASA, https:// www.nasa.gov/kepler/overview/historybyborucki.

10. "About Transits" edited by A. Gould, 2017, NASA, https://www.nasa.gov/kepler /overview/abouttransits.

11. "William Borucki on the Power of Resolve in NASA's Kepler Mission" by I. Evans, 2019, *Elsevier Connect*, https://www.elsevier.com/connect/everyone-agrees -your-method-is-not-going-to-work-william-borucki-on-the-power-of-resolve-in -nasas-kepler-mission.

12. "CoRoT Pictures Transiting Exoplanets" by C. Moutou and M. Deleuil, 2015, *Comptes Rendus Geoscience*, volume 347, 153–158.

13. "List of Large Optical Telescopes," Wikipedia, https://en.wikipedia.org/wiki /List_of_large_optical_telescopes.

14. "Kepler: NASA's First Mission Capable of Finding Earth-Sized Planets," 2009, NASA, https://www.nasa.gov/pdf/314125main_Kepler_presskit_2-19_smfile.pdf.

15. The mission had some problems. The noise was higher than expected, requiring more data and so longer observations to reach the mission goals. In 2012, one of the telescope's four reaction wheels failed. A year later, when a second reaction wheel failed, the telescope was unable to point accurately enough to continue the core mission. A new mission plan, called K2 or "Second Light," was designed; its telescope would acquire less accurate photometry but over a much larger area of sky. The K2 mission extended the productive scientific lifetime of the telescope by four years.

16. "Kepler's Legacy: Discoveries and More," NASA Exoplanet Exploration, https:// exoplanets.nasa.gov/keplerscience/.

17. "Plucking Exoplanets Out of Noisy Kepler Data" by A. Grant, 2021, *Physics Today*, https://physicstoday.scitation.org/do/10.1063/PT.6.1.20210916a/full/.

18. "Kepler Finds 1235 Planets in Four Months: William Borucki Q&A" by A. Haake, 2011, *Popular Mechanics*, https://www.popularmechanics.com/space/telescopes/a6493 /kepler-nasa-telescope-questions-5172767/.

19. "Meet Natalie Batalha, the Explorer Who's Searching for Planets across the Universe" by T. Ferris, 2017, *Smithsonian Magazine*, https://www.smithsonianmag.com /science-nature/natalie-batalha-explorer-searching-planets-across-universe-180967220/.

20. "A Star among Planet Hunters, Natalie Batalha" edited by M. Johnson, 2017, NASA, https://www.nasa.gov/feature/a-star-among-planet-hunters-natalie-batalha.

21. "An Ancient Extrasolar System with Five Sub-Earth-Sized Planets" by T. L. Campante et al., 2015, *Astrophysical Journal*, volume 799, 170–186.

22. "Exoplanet Catalog," 2021, NASA Exoplanet Exploration, https://exoplanets.nasa .gov/discovery/exoplanet-catalog/.

23. "A Revised Exoplanet Yield from the Transiting Exoplanet Survey Satellite (TESS)" by T. Barclay, J. Pepper, and E. V. Quintana, 2018, *Astrophysical Journal Supplement*, volume 239, 2–16.

24. "Space Telescope Delivers the Goods: 2,200 Possible Planets" by P. Brennan, 2021, NASA Exoplanet Exploration, https://exoplanets.nasa.gov/news/1677/space-tele scope-delivers-the-goods-2200-possible-planets/.

Chapter 4

1. "Science Notes" by T. J. Nelson, 2018, https://www.randombio.com/fireflies .html. Randall Monroe also estimates this ratio in an article from his book *What If?* He points out that while fireflies might be feeble in their light output compared to the Sun, they are pound for pound more efficient than the Sun. Bioluminescence is millions of times less efficient than fusion, but fireflies are profligate throughout their short lifetimes while the Sun has to last for billions of years. See https://what-if .xkcd.com/151/.

2. "Observing Exoplanets: What Can We Really See?" by P. Brennan, 2020, NASA Exoplanet Exploration, https://exoplanets.nasa.gov/news/1605/observing-exoplanets -what-can-we-really-see/.

3. "WEIRD: Wide-Orbit Exoplanet Search with Infrared Direct Imaging" by F. Baron et al., 2018, *Astronomical Journal*, volume 156, 137–155.

4. "A Giant Planet Candidate around a Young Brown Dwarf" by G. Chauvin et al., 2004, *Astronomy and Astrophysics*, volume 425, L29–L32.

5. "NASA Exoplanet Archive," 2020, NASA Exoplanet Science Institute, https://exoplanetarchive.ipac.caltech.edu/docs/imaging.html.

6. "A Combined Subaru/VLT/MMT 1–5 Micron Study of Planets Orbiting HR 8799: Implications for Atmospheric Properties, Masses, and Formation" by T. Currie et al., 2011, *Astrophysical Journal*, volume 729, 128–147.

7. "A Low-Mass Planet Candidate Orbiting Proxima Centauri at a Distance of 1.5 AU" by F. Damasso et al., 2020, *Science Advances*, volume 6, eaax7467.

8. "On the Diffraction of an Object-Glass with Circular Aperture" by G. B. Airy, 1835, *Transactions of the Cambridge Philosophical Society*, volume 5, 288–291. The phenomenon was first seen by mathematician John Herschel, but Airy wrote the first theoretical treatment. In optics and astronomy, the pattern is often called an Airy disk. Airy was the astronomer royal of England for nearly fifty years, but his reputation took a hit later in his career when he did not vigorously pursue Neptune and so the discovery went to the French astronomer Urbain Le Verrier. The English and French are rivals in almost everything, astronomy not excluded.

9. "What Is the Diffraction Limit of a Telescope," Cornell University, Astro 201, http://hosting.astro.cornell.edu/academics/courses/astro201/diff_limit.htm.

10. "On the Resolving Power of the Human Eye" by K. N. Ogle, 1951, *Journal of the Optical Society of America*, volume 41, 517–520.

11. Turbulent air is transparent, so it is not obvious why it jumbles up light. The turbulence is caused by convection as hot parcels of air rise and cold parcels of air fall. Hot air is slightly less dense than cold air, and the speed of light varies slightly with the density. This is what causes a uniform and flat wave front of starlight arriving at the top of Earth's atmosphere to be refracted and distorted as it passes through the atmosphere. The refraction angles are small, but cause substantial blurring of the light that has traveled in perfectly straight and parallel lines through the vacuum of space.

12. "The Atmosphere and Observing: A Guide to Astronomical Seeing" by D. Peach, 2005, *Cloudy Nights*, https://www.cloudynights.com/articles/cat/articles/how-to/the-atmosphere-and-observing-a-guide-to-astronomical-seeing-r543.

13. Adaptive optics should not be confused with "active optics." Adaptive optics is the rapid sensing of the wave front distortions of light coming through the atmosphere, followed by rapid changes to a deformable mirror to exactly compensate for those wave front distortions. Active optics is the use of actuators in the cell of a primary telescope mirror to alter its shape to compensate for deformations due to temperature, wind, and mechanical stress as the telescope moves. Adaptive optics operates at ten to a hundred hertz, while active optics is much slower at around one hertz.

14. "The Possibility of Compensating Astronomical Seeing" by H. W. Babcock, 1953, *Publications of the Astronomical Society of the Pacific*, volume 65, 229–236.

15. "The Eye as an Optical Instrument" by P. Artal, in *Optics in Our Time*, edited by M. Al-Amri, M. El-Gomati, and S. Subairy, 2016, London: Springer, 285–297.

16. Wave front sensing uses interferometry in a challenging situation of incoherent white light (most interferometry uses coherent laser light at a single wavelength). First, an optical device turns the aberrations into variations of light intensity. Then a detector transforms the light intensity into an electric signal. Last, a reconstructor uses matrix multiplication to convert the signals into phase aberrations. The phase map is a measure of how much an image is distorted across the focal plane of the telescope. Now you see why I called it a gizmo in the text.

17. "Compensation of Atmospheric Optical Distortion Using a Synthetic Beacon" by C. A. Primmerman, D. V. Murphy, D. A. Page, B. G. Zollars, and H. T. Barclay, 1991, *Nature*, volume 353, 141–143.

18. "Bernard Lyot, 1879–1952," *Astronoo*, http://www.astronoo.com/en/biographies /bernard-lyot.html.

19. For a two-minute animated overview of how a coronagraph works, see "Coronagraph Explanation 2 Minutes," NASA Exoplanet Exploration, https://exoplanets .nasa.gov/resources/2130/coronagraph-explanation-2-minutes/.

20. There was a well-publicized claim of the detection of an exoplanet around the bright star Fomalhaut by the Hubble Space Telescope in 2008. It was independently confirmed and seemed to be a planet twice Jupiter's mass with a seventeen-hundred-Earth-year, highly elliptical orbit at a mean distance of 150 AU. Recent work indicates it is a planetesimal in a dusty disk that has yet to form planets. See "New HST Data and Modeling Reveal a Massive Planetesimal Collision around Fomalhaut" by A. Gaspar and G. H. Rieke, 2020, *Proceedings of the National Academy of Science*, volume 117, 9712–9722.

21. "Starlight Suppression: Technologies for Direct Imaging of Exoplanets" by D. Savransky, 2016, *Frontiers of Engineering: Reports on Leading-Edge Engineering from the 2015 Symposium*, Washington, DC: National Academies Press.

22. "The Woman Who Might Find Us Another Earth" by C. Jones, 2016, *New York Times Magazine*, https://www.nytimes.com/2016/12/07/magazine/the-world-sees-me -as-the-one-who-will-find-another-earth.html.

23. "NASA Wants to Build a Starshade to Hunt Alien Planets" by M. Wall, 2019, Space.com, https://www.space.com/starshade-exoplanet-formation-flying-tech.html. See also "Starshade Would Take Formation Flying to Extremes" by C. Cofield, 2019, NASA Jet Propulsion Laboratory, https://www.jpl.nasa.gov/news/starshade-would-take -formation-flying-to-extremes.

24. "Sarah Seager on Exoplanets, Exploration, and Elusive Earth Twins" by D. Steffens, 2018, *SPIE Magazine*, https://spie.org/news/spie-professional-magazine-archive/2018 -july/star-gazer-sara-seager-on-exoplanets--exploration-and-elusive-earth-twins?SSO=1.

25. "Q&A: Sarah Seager, Exoplanet Explorer" by T. Feder, 2019, *Physics Today*, https://physicstoday.scitation.org/do/10.1063/PT.6.4.20190227a/full/.

26. *Pale Blue Dot: A Vision of the Human future in Space* by C. Sagan, 1997, New York: Ballantine Books.

Chapter 5

1. Details of the fictional world Pandora can be found on the Wiki maintained by fans of James Cameron's movie *Avatar*. See Fandom Avatar Wiki, https://james -camerons-avatar.fandom.com/wiki/Pandora.

2. Alpha Centauri A, the sunlike star that is one of the brightest in the night sky, may have a midsize planet in its habitable zone—a recent claim that is unconfirmed. See "Imaging Low Mass Planets within the Habitable Zone of Alpha Centauri" by K. Wagner et al., 2021, *Nature Communications*, volume 12, https://doi.org/10 .1038/s41467-021-21176-6. Alpha Centauri C, or Proxima Centauri, has two planets: an Earth-size planet in the habitable zone and a super-Earth well outside the habitable zone. See "A Terrestrial Planet Candidate in a Temperate Orbit around Proxima Centauri" by G. Anglada-Escude et al., 2016, *Nature*, volume 536, 437–440. No moon has ever been detected around any star in the Alpha Centauri system.

3. Inevitably, a stigma has been attached to Pandora over the centuries. In fact, Zeus did not punish her because he knew she would open the jar, and curiosity is a virtue in a scientist. She managed to close the jar and contain one evil before it escaped: hopelessness. As a result, humans have hope, and that has tided us through bad times.

4. "Luminous 3-D Jungle Is a Biologist's Dream" by C. K. Yoon, 2010, *New York Times*, https://www.nytimes.com/2010/01/19/science/19essay.html. Before *Avatar*, there was *Return of the Jedi*, the sixth episode of the Star Wars franchise, released in 1983. The film is set on Endor, a forested moon of the gas giant Tana. Endor is home to the Ewok, Dulok, and Yuzzum species, and it is the site of a pivotal battle in the film.

5. "Radar Evidence of Subglacial Liquid Water on Mars" by R. Orosei et al., 2018, *Science*, volume 361, 490–493.

6. "Ocean Worlds Exploration" by J. I. Lunine, 2017, *Acta Astronomica*, volume 131, 123–130.

7. Theory bolsters the expectation that gas giant planets orbiting other stars will have moons similar to what we see in our solar system. Models of moon formation show that they arise naturally as a by-product of the formation of giant planets, with

a combined mass of about 0.01 percent of their planet's mass. See "A Common Mass Scaling for Satellite Systems of Gaseous Planets" by R. M. Canup and W. R. Ward, 2006, *Nature*, volume 441, 834–839.

8. "The HARPS Search for Southern Extra-Solar Planets. XVIII. An Earth-Mass Planet in the GJ 581 Planetary System" by M. Mayor et al., 2009, *Astronomy and Astrophysics*, volume 507, 487–494.

9. "Characterizing the Cool KOIs. III. KOI 961: A Small Star with Large Proper Motion and Three Small Planets" by P. S. Muirhead et al., 2012, *Astrophysical Journal*, volume 747, 144–159.

10. "A Sub-Mercury-Sized Exoplanet" by T. Barclay et al., 2013, *Nature*, volume 494, 452–454.

11. "Why Are Pulsar Planets Rare?" by R. G. Martin, M. Livio, and D. Palaniswamy, 2016, *Astrophysical Journal*, volume 832, 122–130.

12. For example, Kepler can detect a planet like Earth, but the extra dip caused by the Moon would only be 7 percent of Earth's transit dip, which is undetectable. The Moon and Earth orbit a common center of gravity, however, so the Moon tugging on Earth causes a 150-second timing variation in the transit compared to if Earth had no Moon, which is detectable with the best Kepler data.

13. "Exomoon Candidates from Transit Timing Variations: Eight Kepler Systems with TTVs Explainable by Photometrically Unseen Moons" by C. Fox and P. Wiegart, 2020, *Monthly Notices of the Royal Astronomical Society*, volume 501, 2378–2393; "An Independent Analysis of the Six Recently Claimed Exomoon Candidates" by D. Kipping, 2020, *Astrophysical Journal Letters*, volume 900, 44–55.

14. "NASA Supercomputer Assists the Hunt for Exomoons" by S. Lozano, 2017, NASA, https://www.nasa.gov/ames/nasa-supercomputer-assists-the-hunt-for-exomoons.

15. Einstein described his new theory in simple language. See "Space, Time, and Gravitation" by A. Einstein, 1920, *Science*, volume 51, 8–10, https://science.sciencemag.org/content/sci/51/1305/8.full.pdf.

16. "Gravitational Lensing," 2019, NASA Hubble Site, https://hubblesite.org/contents/articles/gravitational-lensing.

17. Microlensing was first seen in 1988 by the amplified light of a distant active galaxy, or quasar. See "Photometric Variations in the Q2237+0305 System: First Detection of a Microlensing Event" by M. J. Irwin, R. L. Webster, P. C. Hewett, R. T. Corrigan, and R. I. Jedrzejewski, 1993, *Astronomical Journal*, volume 98, 1989–1994.

18. "Bodhan Paczynski's Early Years" by G. W. Preston, 2009, in *The Variable Universe: A Celebration of Bodhan Paczynski*, edited by K. Z. Stanek, ASP Conference Series, volume 403, San Francisco: ASP Press, 15–28.

19. Close binary systems lose angular momentum by the release of gravitational waves, altering their evolution. Paczynski predicted this seven years before physicist Russell Hulse and astrophysicist Joseph Taylor detected the effect in a binary pulsar. They were awarded the Nobel Prize for this discovery nineteen years later.

20. "OGLE 2004-BLG-235/MOA-2003-BLG-53: A Planetary Microlensing Event" by I. A. Bond et al., 2004, *Astrophysical Journal Letters*, volume 606, 155–158.

21. Another limitation of microlensing is that the distance to the intervening object, either a star or planet, is only known roughly through a lensing model. That leads to an uncertainty in the mass of the lensing object.

22. "MOA-2011-BLG-262Lb: A Sub-Earth-Mass Moon Orbiting a Gas Giant Primary or a High Velocity Planetary System in the Galactic Bulge" by D. P. Bennett et al., 2014, *Astrophysical Journal*, volume 785, 155–167.

23. "Scientist at Work: Bodhan Paczynski; Finding Opportunity Where Others See Impossibility" by J. Glanz, 1999, *New York Times*, https://www.nytimes.com/1999/01/12/science/scientist-work-bohdan-paczynski-finding-opportunity-where-others-see.html.

24. "Bodhan Pacznyski, 1940–2007, A Biographical Memoir" by B. T. Draine, 2017, National Academy of Sciences, http://www.nasonline.org/publications/biographical-memoirs/memoir-pdfs/paczynski-bohdan.pdf.

Chapter 6

1. "Interview: Michel Mayor, Nobel Prize-winning Astrophysicist" by H. Lamb, 2020, Engineering and Techology, https://eandt.theiet.org/content/articles/2020/08/interview-michel-mayor-nobel-prize-winning-astrophysicist/.

2. In genetics, the cost of sequencing a genome has gone down from a hundred million dollars in 2001 for the first to a thousand dollars per genome as millions are sequenced. In high-energy physics, improvements in superconducting magnets led to the hundredfold increase in the collision energy of the Large Hadron Collider. In astronomy, CCDs of the late 1970s had a million pixels while the camera for the Very Rubin Observatory in Chile will have three billion pixels. Across all fields in science, exponential gains in computing speed and storage capacity have spurred progress.

3. Jessie Christiansen, Twitter, https://twitter.com/aussiastronomer/status/790700485376348160?lang=en.

4. NASA Exoplanet Science Institute, Staff, http://web.ipac.caltech.edu/staff/christia/; quote from Griffith University Outstanding Young Alumnus Award video, 2018, https://www.youtube.com/watch?v=LsY7WSIqmmQ.

5. "Another Day, Another Exoplanet, and Scientists Just Can't Keep Up" by M. Bartels, 2019, Space.com, https://www.space.com/so-many-exoplanets-discovery-pace.html.

6. *Understanding Moore's Law: Four Decades of Innovation* edited by D. C. Brock, 2006, Philadelphia: Chemical Heritage Foundation.

7. NASA Exoplanet Exploration, Discoveries Dashboard, https://exoplanets.nasa.gov /discovery/discoveries-dashboard/.

8. "We're Not Prepared for the End of Moore's Law" by D. Rotman, 2020, *MIT Technology Review*, https://www.technologyreview.com/2020/02/24/905789/were-not -prepared-for-the-end-of-moores-law/.

9. "Exoplanet Vision 2050" by R. Heller and L. L. Kiss, 2019, based on an invited talk at the Exoplanet Vision 2050 Workshop at Konkoly Observatory, https://arxiv .org/abs/1911.12114. For a more lighthearted view of the dangers of extrapolation, see the "The Existential Threat of Future Exoplanet Discoveries" by M. B. Lund, 2021, https://arxiv.org/pdf/2103.17079.pdf.

10. "Transiting Exoplanet Survey Satellite" by G. R. Ricker et al., 2014, *Journal of Astronomical Telescopes, Instruments, and Systems*, volume 1, 014003.

11. "New Explorer Mission Chooses the Just-Right Orbit," NASA, https://www.nasa .gov/content/goddard/new-explorer-mission-chooses-the-just-right-orbit. The trick works because TESS has an orbital period of 13.7 days, which puts it in a 2:1 resonance with the Moon. The stable orbit means little fuel is required for thrusters to maintain its position, which is critical in a low-mass spacecraft like TESS.

12. "The First Habitable-Zone Earth-Sized Planet from TESS. I. Validation of the TOI-700 System" by E. A. Gilbert, 2020, *Astronomical Journal*, volume 160, 116–136.

13. "What's Next for Exoplanet Searches? Live Science Talks with Astrophysicist Sara Seager" by Y. Saplakoglu, 2018, *Live Science*, https://www.livescience.com /62572-sara-seager-genius-gala-exoplanets.html.

14. The measurement of positions on the sky, in the form of two angles, is called "astrometry." For a history of this rich subject, which impacts the study of every type of astronomical object, see "The History of Astrometry" by M. Perryman, 2012, *European Physics Journal*, volume 37, 745–792.

15. *Astronomical Applications of Astrometry: Ten Years of Exploitation of the Hipparcos Satellite Data* by M. Perryman, 2008, Cambridge: Cambridge University Press.

16. European Space Agency, Exoplanets, 2019, https://sci.esa.int/web/gaia/-/58784 -exoplanets. The accurate photometry of Gaia will also be sensitive to transits, but the spectrograph is not accurate enough to detect the Doppler wobble.

17. "Astrometric Exoplanet Detection with *Gaia*" by M. Perryman, J. Hartman, G. A. Bakos, and L. Lindegren, 2014, *Astrophysical Journal*, volume 797, 14–35.

18. "The PLATO 2.0 Mission" by H. Rauer et al., 2014, *Experimental Astronomy*, volume 38, 249–330.

19. PLATO is a medium-class mission in ESA's long-range plan, called *Cosmic Vision: Space Science for Europe 2015–2025*. Other missions relevant to exoplanets and astrobiology are the small mission CHEOPS, designed to measure exoplanet sizes and with a science team led by Didier Queloz; ARIEL, a medium-size mission to observe exoplanet atmospheres in the infrared; and JUICE, a mission to study three of Jupiter's moons thought to have subsurface bodies of water—Ganymede, Callisto, and Europa.

20. "PLATO Science Goals," European Space Agency, https://sci.esa.int/web/plato /-/42277-science.

21. The Nancy Grace Roman Space Telescope, or Roman Space Telescope, started life as the SNAP dark energy mission in 2008. It then evolved into the DOE-NASA Joint Dark Energy Mission in 2010. This got merged with two other concepts in the astronomy Decadal Survey to become the Wide Field Infrared Survey Telescope (WFIRST), one of the blizzard of NASA acronyms that can dull the senses. After it got upgraded to a larger mirror and its funding became more secure in 2020, it was renamed after Nancy Grace Roman, the first woman to serve as the chief of astronomy at NASA, and someone considered the "mother" of the Hubble Space Telescope. See https://www.nasa.gov/press-release/nasa-telescope-named-for-mother-of -hubble-nancy-grace-roman.

22. "The Telescope That Ate Astronomy" by L. Billings, 2010, *Nature*, volume 467, 1028–1030.

23. "Ex-Spy Telescope May Get New Identity as Space Investigator" by D. Overbye, 2012, *New York Times*, https://www.nytimes.com/2012/06/05/science/space /repurposed-telescope-may-explore-secrets-of-dark-energy.html.

24. "Doppler Spectroscopy as a Path to the Detection of Earth-Like Planets" by M. Mayor, 2014, *Nature*, volume 513, 328–335.

25. "Radial Velocity Jitter of Stars as a Functional of Observational Timescale and Stellar Age" by S. S. Brems, M. Kurster, T. Trifonov, S. Reffert, and A. Quirrenbach, 2019, *Astronomy and Astrophysics*, volume 632, A37–A51.

26. Mayor's first spectrograph was called COREVAL, and it operated at the 1.9-meter telescope at Haute-Provence Observatory. Its successor, used in the same telescope, led to the discovery of the first exoplanet in 1995. Then Mayor led the development of HARPS, a spectrograph in Chile on the European Southern Observatory's 3.6-meter telescope, which discovered 130 exoplanets starting in 2002. The new state-of-the art spectrograph is called ESPRESSO, and it operates on the 8.4-meter Very Large Telescope in Chile. You can see that the progression to higher precision also involved larger and larger telescopes. For information on ESPRESSO,

see "ESPRESSO at VLT. On-Sky Performance and First Results" by F. Pepe et al., 2021, *Astronomy and Astrophysics*, volume 645, A96–A121.

27. "The Crucial Role of Ground-Based Doppler Measurements for the Future of Exoplanet Science" by J. H. Steffen et al., 2018, white paper submitted to NAS Exoplanet Strategy Committee, https://arxiv.org/abs/1803.06057.

28. "Which Stars Can See Earth as a Transiting Exoplanet?" by L. Kaltenegger and J. Pepper, 2020, *Monthly Notices of the Royal Astronomical Society*, volume 499, L111–L115. Also, "Past, Present, and Future Stars That Can See Earth as a Transiting Exoplanet" by L. Kaltenegger and J. K. Faherty, 2021, *Nature*, volume 594, 505–507.

29. "Astronomer Identify Closest Stars That Could See Earth as a Transiting Exoplanet," 2020, *Sci-News*, http://www.sci-news.com/astronomy/earth-transiting-exoplanet-08978 .html.

Chapter 7

1. There is a vast literature, and a mixture of science and art, used to decide which lures work best for which species of fish. This complexity is vastly simplified to make the analogy with exoplanet detection. For a sense of this literature, consider the monumental work of Robert Slade, who wrote a nineteen-volume *Encyclopedia of Old Fishing Lures Made in North America*, 2011, Victoria, Canada: Trafford Publishing. Each book runs 450 pages, making it over 8,000 pages and 10,000 pictures covering the esoterica of fishing lures.

2. For an explanation of the technology of sonar as it is used to detect fish, see https://dosits.org/people-and-sound/fishing/how-is-sound-used-to-identify-fish/. Another important detail is the sonar is not able to image fish effectively within the top meter of a body of water due to an effect called "surface clutter." For an explanation, see https://deepersonar.com/us/en_us/how-it-works/how-sonars-work.

3. *The Hunting of the Snark* by L. Carroll, 1876, London: Macmillan Publishers.

4. The motion is called a "reflex motion." Two children with equal weight on a teeter-totter or seesaw would make it balance. An adult three times the child's weight could balance the teeter-totter by sitting one-third of the way out from the pivot point. With stars, two equal mass stars in a binary system orbit around a point halfway between them. As the masses become more unequal, the center of motion moves closer to the more massive object. With planets, the center of motion is usually within the star, so it pirouettes around its edge, but not at its center.

5. Michel Mayor and Didier Queloz discovered the planet orbiting 51 Peg with the 1.9-meter telescope at Haute Provence Observatory in southern France. There are dozens of telescopes 4-meters or larger, so this is a modest-size telescope by modern standards.

6. Formally, the observed Doppler velocity is V x sin(i), where V is the true velocity of the star, and i is the orbit inclination angle to the line of sight. The inferred mass of the planet is reduced by that same factor of sin(i). The average of sin(i) over 4 π steradians is 0.5, so that is the factor by which planet mass is on average underestimated.

7. "Exploring Exoplanet Populations with NASA's Kepler Telescope" by N. N. Batalha, 2014, *Proceedings of the National Academy of Sciences*, volume 111, 12647–12654.

8. Proxima Centauri is a red dwarf, seven times smaller than the Sun. Yet even the Sun at the same distance, with an angle of 0.007 arc seconds, would be too small to measure with conventional methods.

9. "How Large Can a Planet Be?" by B. Koberlein, 2019, Phys.org, https://phys.org /news/2019-11-large-planet.html.

10. You could say that it's easy to distinguish between stars and planets, because stars form the way stars do and planets form the way planets do. But the observations don't reveal how a gas giant planet or brown dwarf formed. As an example of the ambiguity, see "Exoplanet Discovery Blurs the Line between Large Planets and Small Stars" by A. Norton, 2019, Phys.org, https://phys.org/news/2019 -09-exoplanet-discovery-blurs-line-large.html.

11. "The Binary Star 70 Ophiuchi Revisited" by W. D. Heintz, 1988, *Journal of the Royal Astronomical Society of Canada*, volume 82, 140–145.

12. "A Mass for the Extrasolar Planet Gliese 876b Determined from Hubble Space Telescope Fine Guidance Sensor 3 Astrometry and High Precision Radial Velocities" by G. F. Benedict et al., 2002, *Astrophysical Journal Letters*, volume 581, 115–118.

13. "Astrometric Orbit of a Low-Mass Companion to an Ultra-Cool Dwarf" by J. Sahlmann, P. F. Lazorenko, D. Segransan, E. L. Martin, D. Queloz, M. Mayor, and S. Udry, 2013, *Astronomy and Astrophysics*, volume 553, A133–A142.

14. An exoplanet in the galaxy M51, twenty-three million light-years away, was found using the novel technique of an X-ray transit. The exoplanet is a gas giant orbiting a neutron star or black hole in a binary system with a massive star. See N. Whitehead and M. Watzke, "Astronomers May Have Discovered the First Planet Outside of Our Galaxy," 2021, Harvard Center for Astrophysics, https://www.cfa .harvard.edu/news/astronomers-may-have-discovered-first-planet-outside-our-galaxy.

15. "At Least One in Six Stars Has an Earth-Sized Planet" by D. A. Aguillar and C. Pulliam, 2013, Harvard-Smithsonian Center for Astrophysics, https://www.cfa .harvard.edu/news/2013-01.

16. These fractions are probably underestimated because at the time Kepler data had not been taken long enough to find planets with multiyear orbital periods. These planets are relatively hot. For example, the Earth fraction is of planets

orbiting in 85 days or less, so closer than Mercury is to the Sun. The fraction of gas giants is for those that are orbiting much closer to their stars than the gas giants in our solar system.

17. "One or More Bound Planets per Milky Way Star from Microlensing Observations" by A. Cassan et al., 2011, *Nature*, volume 481, 167–169.

18. "The Occurrence of Rocky Habitable Zone Planets around Solar-Like Stars from Kepler Data" by S. Bryson et al., 2020, *Astronomical Journal*, volume 161, https://iopscience.iop.org/article/10.3847/1538-3881/abc418.

19. "Exploring Exoplanet Populations with NASA's Kepler Telescope" by N. N. Batalha, 2014, *Proceedings of the National Academy of Sciences*, volume 111, 12647–12654.

20. "Are We Missing Other Earths?" by K. Lester, S. Howell and A. Kocz, 2021, NOIRLab, https://noirlab.edu/public/news/noirlab2120/.

21. Quoted by Margaret Fuller in *The Varieties of Religious Experience* by W. James, 1902, New York: Modern Library, 41.

Chapter 8

1. *Troilus and Cressida* by W. Shakespeare, 1609, act 1, scene 3.

2. "How Was the Sun Formed?" by N. T. Redd, 2017, Space.com, https://www.space.com/19321-sun-formation.html.

3. In any scenario, exceptions are allowed, as long as they are rare and do not break the rules too badly. Venus is the only planet that spins in the opposite sense to all the others. In physical terms, its angular momentum is not aligned with that of the Sun and all the other planets. This is assumed to have been caused by a collision early on, during the violent early phase of planet building. No other evidence of the collision would have survived after so long. A similar collision involving Earth is thought to have led to the formation of the Moon, which is unusually large for a moon of a terrestrial planet.

4. *Emanuel Swedenborg, Scientist and Mystic* by S. Toksvig, 1948, New Haven, CT: Yale University Press.

5. "Kant's Cosmogony Re-Evaluated" by S. Palmquist, 1987, *Studies in History and Philosophy of Science*, volume 18, 255–269.

6. "Leonard Medal Citation for Victor Sergeivitch Safronov" by G. W. Wetherill, 1989, *Meteoritics*, volume 24, 347.

7. *Just So Stories* by Rudyard Kipling was a fanciful set of stories, published in 1902, that pretended to explain the characteristics of animals, such as the spots on a

leopard and long trunk of an elephant. In science and philosophy, the term refers to an ad hoc fallacy, an unverifiable and unfalsifiable explanation.

8. "Gas Giant Interiors," 2003, NASA/Lunar and Planetary Institute, https://solarsystem .nasa.gov/resources/677/gas-giant-interiors-2003/.

9. The roughly geometric spacing was named the Titius-Bode law, after the eighteenth-century astronomers who called attention to it. The law got a boost in 1781 with the discovery of Uranus, and then in 1800 with the discovery of Ceres, both of which fit the progression. But then it was weakened when Neptune and Pluto did not fit the pattern, and Ceres was found to be one of many asteroids and not a planet. There is no physical explanation for the geometric spacing of planets, so it is now considered more of a coincidence than a law of nature. Astronomers have tried to fit the progression to exoplanets, with some success. See "Exoplanet Predictions Based on the Generalized Titius-Bode Relation" by T. Bovaird and C. H. Lineweaver, 2013, *Monthly Notices of the Royal Astronomical Society*, volume 435, 1126–1138.

10. "Origins of Hot Jupiters" by R. I. Dawson and J. A. Johnson, 2018, *Annual Reviews of Astronomy and Astrophysics*, volume 56, 175–221.

11. "Why Hot Jupiter Exoplanets Aren't Eaten by Their Stars" by M. Wall, 2013, Space.com, https://www.space.com/21473-alien-planets-migration-hot-jupiters.html.

12. "K2–137b: An Earth-Sized Planet in a 4.3-Hour Orbit around an M-Dwarf" by A. M. S. Smooth et al., 2018, *Monthly Notices of the Royal Astronomical Society*, volume 474, 5523–5533.

13. "Interview with Alan Boss," in *Talking about Life: Conversations on Astrobiology*, edited by C. D. Impey, 2010, Cambridge: Cambridge University Press, 248.

14. Spectroscopy has been a primary tool of astronomy since Isaac Newton used a prism to spread sunlight into its constituent colors. Two hundred years ago, physicist Joseph von Fraunhofer observed narrow absorption lines in the Sun's spectrum, and this information eventually revealed that the Sun is made primarily of hydrogen and helium. Early in the twentieth century, spectroscopy led to an understanding of the chemical composition, physical structure, and evolution of all stars. Now with exoplanets, spectroscopy is a state-of-the-art tool for learning more about their composition.

15. The star's absorption spectrum must be measured accurately to be able to subtract it and see the small amount of absorption imprinted by the exoplanet atmosphere. These observations are typically done from the stable environment of a space telescope. For details on the method, see "Exoplanet Atmosphere Measurements from Transmission Spectroscopy and Other Planet-Star Combined Light Observations" by L. Kreidberg, 2018, https://arxiv.org/pdf/1709.05941.pdf.

16. "A Spectrum of an Extrasolar Planet" by L. J. Richardson, D. Deming, K. Horning, S. Seager, and J. Harrington, 2007, *Nature*, volume 445, 892–895. This paper

reported the infrared detection of silicate clouds. Another investigator found water (steam, at this high temperature) in the atmosphere. A third study found sodium and vanadium oxide, but not water. As a sign of the difficulty of the observations and interpretation, and controversy in this field, a fourth study in 2021 did not see any evidence for a planetary atmosphere. Recently, astronomers sniffed the atmosphere of a hot Jupiter and showed from its chemistry that it must have migrated inward from farther out in the system. See "First Transiting Exoplanet's Chemical Fingerprint Reveals Its Distant Birthplace," 2021, *Science Daily*, https://www.sciencedaily .com/releases/2021/04/210407114208.htm.

17. "Aerosol Composition of Hot Giant Exoplanets Dominated by Silicates and Hydrocarbon Hazes" by P. Gao et al., 2020, *Nature Astronomy*, volume 4, 951–956.

18. "Rains of Terror on Exoplanet HD 189733b" by S. Loff, 2017, NASA, https:// www.nasa.gov/image-feature/rains-of-terror-on-exoplanet-hd-189733b.

19. "Nightside Condensation of Iron in an Ultrahot Giant Planet" by D. Ehrenreich et al., 2020, *Nature*, volume 580, 597–601.

20. "Exoplanet Exploration: Planets beyond Our Solar System," 2021, NASA, https://exoplanets.nasa.gov/what-is-an-exoplanet/planet-types/overview/.

21. "Particles, Environments, and Possible Ecologies in the Jovian Atmosphere" by C. Sagan and E. E. Salpeter, 1976, *Astrophysical Journal Supplement*, volume 32, 737–755.

22. "How Stratospheric Life Is Teaching Us about the Possibility of Extreme Life on Other Worlds" by S. Vartan, 2018, Astrobiology at NASA, https://astrobiology.nasa .gov/news/how-stratospheric-life-is-teaching-us-about-the-possibility-of-extreme -life-on-other-worlds/.

23. "Edwin Salpeter and the Gasbags of Jupiter" by P. Gilster, 2009, *Centauri Dreams* (blog), https://www.centauri-dreams.org/2009/02/25/edwin-salpeter-and-the-gasbags -of-jupiter/.

Chapter 9

1. *The Greek and Roman Myths: A Guide to the Classical Stories* by P. Matyszak, 2010, London: Thames and Hudson.

2. *The Merchant of Venice* by W. Shakespeare, 1619, second quarto, act 5, scene 1.

3. "The Harmony of the Spheres" by B. R. Gaizauskas, 1974, *Journal of the Royal Astronomical Society of Canada*, volume 68, 146–151.

4. *The Harmony of the World* by J. Kepler, 1609, translated by E. J. Aiton, A. M. Duncan, and J. V. Field, 1997, Philadelphia: American Philosophical Society. Kepler went beyond simply looking for numerical ratios in the planet orbits. He tried to fit the five perfect or Pythagorean solids into the spacing of the six orbits. The solids

and their number of identical faces are the tetrahedron (four), cube (six), octahedron (eight), dodecahedron (twelve), and icosahedron (twenty).

5. *History of the Planetary Systems from Thales to Kepler* by J. L. E. Dreyer, 1906, Cambridge: Cambridge University Press.

6. "A Freaky Fluid Inside Jupiter?" by D. Coulter, 2011, NASA Science, https://science.nasa.gov/science-news/science-at-nasa/2011/09aug_juno3.

7. "Researchers Recreate the Ice Giants' Interiors" by N. T. Redd, 2018, *Astronomy*, https://astronomy.com/news/2018/04/researchers-recreate-the-ice-giants-interiors.

8. Exoplanet research is still too immature to answer the question, and the observational biases that affect the exoplanets that can and cannot be detected are still persistent. Theorist Sean Raymond has tried to answer the question based only on the relationship of Jupiter to the Sun. His response is that we are one in two thousand. "Exactly How Unusual Is Our Solar System?" by S. Raymond, 2016, *PlanetPlanet* (blog), https://planetplanet.net/2016/07/12/exactly-how-unusual-is-our-solar-system/.

9. "Orbital Resonance," 2021, Space Wiki, https://space.fandom.com/wiki/Orbital_resonance#Types_of_resonance.

10. Resonance is familiar with physical objects too. A string or hollow tube has a natural frequency, and that sound can be induced by a nearby object vibrating or oscillating at that frequency. The natural frequency also has higher harmonics or overtones that are related to the natural frequency by the ratio of whole numbers, and they can be induced by vibrations of a nearby object as well. In one practical example, soldiers are usually ordered to break step when marching over a bridge, in case the rhythm induces a resonance in the bridge structure, with potentially catastrophic effects!

11. "Origin of the Orbital Architecture of the Giant Planets of the Solar System" by K. Tsiganis, R. Gomes, A. Morbidelli, and H. F. Levison, 2005, *Nature*, volume 435, 459–461.

12. The process is governed by the physical laws of the conservation of energy and conservation of angular momentum. The small icy rocks lose energy and angular momentum with each encounter, while the giant planets gain energy and angular momentum, thus moving to a larger orbit. The total energy and angular momentum of the solar system is unchanged, but it gets redistributed through the interactions.

13. "The Original Nice Model of Planet Migration: The Gas and Ice Giants" by S. Stanley, 2020, https://www.thegreatcoursesdaily.com/the-original-nice-model-of-planet-migration-the-gas-and-ice-giants/.

14. The full story is more complex than can be described here, and the Nice model had shortcomings that have been addressed since 2005. One major adjustment has Jupiter with a more active role, forming at 3.5 AU, then migrating inward to the distance of Mars before reversing course and heading out to its current position.

This is called the Grand Tack model, with reference to a sailboat rapidly changing course. "Outward Migration of Jupiter and Saturn in 3:2 or 2:1 Resonance in Radiative Disks: Implications for the Grand Tack and Nice Models" by A. Pierens et al., 2014, *Astrophysical Journal Letters*, volume 795, L11–L17.

15. "Solar System Formation in the Context of Extrasolar Planets" by S. N. Raymond, A. Izodoro, and A. Morbidelli, 2020, in *Planetary Astrobiology*, edited by V. Meadows, G. Arney, B. Schmidt, and D. J. Des Marais, Tucson: University of Arizona Press, 287–324.

16. "Rare Chance to Reach Ice Giants Excites Scientists" by E. Gibney, 2020, *Nature*, volume 579, 17.

17. "Most Common Exoplanets Are Weird Mini-Neptunes" by V. Jaggard, 2014, *New Scientist*, https://www.newscientist.com/article/dn24826-most-common-exoplanets-are-weird-mini-neptunes/.

18. "Identifying Exoplanets with Deep Learning: A Five-Planet Resonant Chain around Kepler-80 and an Eighth Planet around Kepler-90" by C. J. Shallue and A. Vanderburg, 2018, *Astronomical Journal*, volume 155, 94–114.

19. "Puzzling Six-Exoplanet System Found with Resonant Rhythmic Movement," 2021, *Sci Tech Daily*, https://scitechdaily.com/puzzling-six-exoplanet-system-discovered-with-resonant-rhythmic-movement/.

20. *Sounds of Exoplanet Systems* by M. Russo, 2019, YouTube, https://www.youtube.com/watch?v=idlB8JgKGU4.

21. "Set to Music, Exoplanets Reveal Insights on Their Formation" by N. T. Redd, 2019, *Eos: Science News by AGU*, https://eos.org/articles/set-to-music-exoplanets-reveal-insight-on-their-formation.

22. "The California-Kepler Survey. III. A Gap in the Radius Distribution of Small Planets" by B. J. Fulton et al., 2017, *Astronomical Journal*, volume 154, 109–128.

23. "New Branch in Family Tree of Planets Discovered" by M.-E. Naud, 2017, Institute for Research on Exoplanets, http://www.exoplanetes.umontreal.ca/new-branch-in-family-tree-of-exoplanets-discovered/?lang=en.

Chapter 10

1. *The Comedy of Errors* by W. Shakespeare, 1595, act 1, scene 2.

2. "Biology and Pathology of Water" by A. Szent-Györgyi, 1971, *Perspectives on Biology and Medicine*, volume 14, 239–249.

3. "Biological Roles of Water: Why Is Water Essential for Life?" by M. Sargen, 2019, *Science in the News* (blog), Harvard University, https://sitn.hms.harvard.edu/uncategorized/2019/biological-roles-of-water-why-is-water-necessary-for-life/.

4. "Water and Life: The Medium Is the Message" by M. Frenkel-Pinter, V. Rajaei, J. B. Glass, N. V. Hud, and L. D. Williams, 2021, *Journal of Molecular Evolution*, volume 89, 2–11.

5. "Life Can Survive on Much Less Water Than You Think" by A. Hadhazy, 2014, *Astrobiology Magazine*, https://www.astrobio.net/extreme-life/life-can-survive-much -less-water-might-think/.

6. "X-Treme Microbes: Amazing Survivors," 2018, National Science Foundation, https://www.nsf.gov/news/special_reports/microbes/images/Xtreme_microbes_sur vivors_summary.pdf.

7. "Tardigrades Return from the Dead" by J. Fox-Skelly, 2015, BBC, http://www.bbc .com/earth/story/20150313-the-toughest-animals-on-earth.

8. "Biological Constraints on Habitability" by L. Dartnell, 2011, *Astronomy and Geophysics*, volume 52, 25–28.

9. "Radioactivity May Fuel Life Deep Underground and Inside Other Worlds" by J. Cepelewicz, 2021, *Quanta Magazine*, https://www.quantamagazine.org/radioactivity -may-fuel-life-deep-underground-and-inside-other-worlds-20210524/.

10. "Following the Water, the New Program for Mars Exploration" by G. S. Hubbard, F. M. Naderi, and J. B. Garvin, 2002, *Acta Astronautica*, volume 51, 337–350.

11. "Earth-Like Planets Probably Water-logged" by P. Ball, 2003, *Nature News*, https://www.nature.com/news/2003/030818/full/news030818-10.html.

12. *Black Holes, Stars, Earth and Mars: Astronomy Poems for All Ages* by S. Raymond, 2020, independent publisher.

13. "Build a Better Solar System" by S. Raymond, 2013, *PlanetPlanet* (blog), https:// planetplanet.net/2013/06/11/build-a-better-solar-system/.

14. "1 Million Habitable Planets Could (Theoretically) Orbit a Black Hole. Here's How" by C. Q. Choi, 2018, Space.com, https://www.space.com/40846-black-hole -million-habitable-planets.html.

15. There may not be a division or sharp transition between mini-Neptunes and super-Earths. They may exist on a continuum of structures and compositions. See "Irradiated Ocean Planets Bridge Super-Earth and Sub-Neptune Populations" by O. Mousis et al., 2020, *Astrophysical Journal Letters*, volume 896, L22–L26.

16. "A Super-Earth Transiting a Low-Mass Star" by D. Charbonneau et al., 2009, *Nature*, volume 462, 891–894.

17. In modeling exoplanet structures, this ambiguity is called a degeneracy in the models. For example, taking an artificial case of a planet with two layers, it's possible to dream up a planet with a small metallic core and large gaseous envelope

that has the same mean density as a planet that is mostly water with a thin outer atmosphere. As long as planet hunters only have mass (from radial velocity measurements) and size (from transits), little can be deduced about exoplanet structure.

18. "The Skies of Mini-Neptunes" by H. Wakeford, 2015, *Planetary Report*, https://www.planetary.org/articles/the-skies-of-mini-neptunes.

19. "How Long Can Humans Survive without Food or Water?" by T. Puiu, 2020, *ZME Science*, https://www.zmescience.com/other/feature-post/how-long-survive-no-food-water-052352/.

20. "Meet Aeroice: Scientists Just Engineered the Lowest Density Ice Crystals Ever" by P. Dockrill, 2017, *Science Alert*, https://www.sciencealert.com/scientists-just-discovered-a-whole-new-family-of-ice-phases.

21. "Scientists Just Created a Bizarre Form of Ice That's Half as Hot as the Sun" by M. Starr, 2019, *Science Alert*, https://www.sciencealert.com/exotic-form-of-water-ice-is-half-as-hot-as-the-sun-and-it-s-just-been-created-here-on-earth.

22. *Cat's Cradle* is full of dark humor and clever jabs at scientists and their occasional hubris. Vonnegut talked about the idea of Ice Nine with his brother, who had a PhD in physical chemistry from MIT. He was also influenced by a story told at General Electric, where he worked for a while in public relations. In the 1930s, H. G. Wells had visited GE and got the idea of water that was solid at room temperature from Nobel laureate chemist Irving Langmuir. Wells never used the idea in his own writing.

23. "Weird Water Phase 'Ice-VII' Can Grow as Fast as 1,000 Miles per Hour" by J. Ouellette, 2018, *Ars Technica*, https://arstechnica.com/science/2018/10/weird-water-phase-ice-vii-can-grow-as-fast-as-1000-miles-per-hour/.

24. "Unfathomably Deep Oceans on Alien Water Worlds" by P. S. Anderson, 2019, *Earth Sky*, https://earthsky.org/space/exoplanet-water-worlds-deep-oceans-2019-study.

25. "Beneath the Surface of Our Galaxy's Water Worlds" by A. Salles, 2020, Argonne National Laboratory, https://www.anl.gov/article/beneath-the-surface-of-our-galaxys-water-worlds.

26. For a nontechnical summary of the topic, see "Are Water Worlds Habitable?" by S. Hall, 2018, *Scientific American*, https://www.scientificamerican.com/article/are-water-worlds-habitable/. For a technical article that gives a more positive prognosis about habitability based on simulations, see "Habitability of Exoplanet Waterworlds" by E. S. Kite and E. B. Ford, 2018, *Astrophysical Journal*, volume 864, 75–101.

27. "Introducing Hycean Planets," by M. Kaufman, 2021, Many Worlds, https://manyworlds.space/2021/08/31/introducing-hycean-planets/.

28. "Habitability and Biosignatures of Hycean Worlds" by N. Manhusudhan, A. A. A. Piette, and S. Constantinou, 2021, *Astrophysical Journal*, volume 918, 1–26.

29. "Many Chemistries Could Be Used to Build Living Systems" by W. Bains, 2004, *Astrobiology*, volume 4, 137–157.

30. "Evaluating Alternatives to Water as Solvents for Life: The Example of Sulfuric Acid" by W. Bains, J. J. Petkowski, Z. Zhan, and S. Seager, 2021, *Life*, volume 11, 400–433.

31. "Diving into Exoplanets: Are Water Seas the Most Common?" by F. J. Balleseros, A. Fernandez-Soto, and V. J. Martinez, 2019, *Astrobiology*, volume 19, 642–654.

Chapter 11

1. *The Comedy of Errors* by W. Shakespeare, 1595, act 5, scene 1.

2. "Discoveries Dashboard," NASA Exoplanet Exploration, https://exoplanets.nasa .gov/discovery/discoveries-dashboard/.

3. "Red Dwarfs: The Most Common and Longest-Lived Stars" by N. T. Redd, 2019, Space.com, https://www.space.com/23772-red-dwarf-stars.html.

4. "The Potential of Planets Orbiting Red Dwarf Stars to Support Oxygenic Photo-synthesis and Complex Life" by J. Gale and A. Wandel, 2017, *International Journal of Astrobiology*, volume 9, 1–16.

5. "Billions of Rocky Planets in the Habitable Zones around Red Dwarfs," 2015, Instituto Nazionale di Astrofisica, http://www.inaf.it/en/inaf-news/billions-of-rocky -planets-in-the-habitable-zones-around-red-dwarfs.

6. "Seven Temperate Terrestrial Planets around the Nearby Ultra-Cool Dwarf Star TRAPPIST-1" by M. Gillon et al., 2017, *Nature*, volume 542, 456–460.

7. "Update on the 7 Earth-Size Planets Orbiting Nearby TRAPPIST-1" by A. Oliver, 2021, *Earth Sky*, https://earthsky.org/space/7-trappist-1-planets-similar-composition -unlike-earth.

8. "NASA Planet Hunter Finds Earth-Sized Habitable-Zone World" by J. Kazmier-czak, 2020, NASA Exoplanet Exploration, https://exoplanets.nasa.gov/news/1617 /nasa-planet-hunter-finds-earth-size-habitable-zone-world//.

9. "Space Telescope Delivers the Goods: 2,200 Possible Exoplanets" by P. Brennan, 2021, NASA Exoplanet Exploration, https://exoplanets.nasa.gov/news/1677/space -telescope-delivers-the-goods-2200-possible-planets/. For the technical paper describ-ing the candidates, see "The TESS Objects of Interest Catalog from the TESS Primary Mission" by N. M. Guerrero, 2021, https://arxiv.org/abs/2103.12538.

10. "Finding Another Earth" by P. Brennan, 2015, NASA, https://www.nasa.gov/jpl /finding-another-earth.

11. "SETI Targets Kepler 425b, Earth's Cousin, in Search for Alien Life" by N. T. Redd, 2015, Space.com, https://www.space.com/30114-seti-alien-life-kepler-452b -earth-cousin.html.

12. "A Terrestrial Planet Candidate in a Temperate Climate around Proxima Centauri" by G. Anglada-Escude, 2016, *Nature*, volume 536, 437–440. Dampening the excitement is the observation of an enormous flare from the red dwarf, a hundred times more powerful than any similar event seen on the Sun. Life on this planet might be challenged unless it is protected by a thick atmosphere. See "Enormous Flare from Sun's Nearest Neighbor Breaks Records," 2021, *Science Daily*, https://www .sciencedaily.com/releases/2021/04/210421124519.htm.

13. "Revisiting Proxima with ESPRESSO" by A. S. Mascareno et al., 2020, *Astronomy and Astrophysics*, volume 639, A77.

14. "Imaging Low-Mass Planets within the Habitable Zone of Alpha Centauri" by K. Wagner et al., 2021, *Nature Communications*, volume 12, 922.

15. "Scientists Plan Private Mission to Hunt for Earths around Alpha Centauri" by J. O'Callaghan, 2021, *Scientific American*, https://www.scientificamerican.com/article /scientists-plan-private-mission-to-hunt-for-earths-around-alpha-centauri/.

16. "A Two-Tiered Approach to Assessing the Habitability of Exoplanets" by D. Schulze-Makuch et al., 2011, *Astrobiology*, volume 11, 1041–1052.

17. "Habitable Exoplanets Catalog: Data of Potentially Habitable Worlds," Planetary Habitability Laboratory, University of Puerto Rico at Arecibo, http://phl.upr .edu/projects/habitable-exoplanets-catalog/data.

18. The general rationale for this statement is the principle of mediocrity, or the Copernican principle. It assumes that there is nothing intrinsically unlikely or unusual about Earth's formation and cosmic location.

19. "Terrestrial Planets across Space and Time" by E. Zackrisson, P. Calissendorff, J. Gonzales, A. Benson, A. Johansen, and M. Janson, 2016, *Astrophysical Journal*, volume 833, 214–225.

20. "Parallel Universes" by M. Tegmark, 2003, *Scientific American*, May, 41–51.

21. "Patterns in Paleontology: The Latitudinal Biodiversity Gradient" by D. P. Mannion, P. Upchurch, D. P. Benson, and A. Goswami, 2014, *Paleontology Online*, volume 4, 1–8.

22. "Superhabitable Worlds" by R. Heller and J. Armstrong, 2014, *Astrobiology*, volume 14, 50–66.

23. "In Search of a Planet Better Than the Earth: Top Contenders for a Superhabitable World" by D. Schulze-Makuch, R. Heller, and E. Guinan, 2020, *Astrobiology*, volume 20, 1394–1404.

24. "Ingenious: Lisa Kaltenegger" by Michael Segal, 2013, *Nautilus*, https://nautil.us /ingenious-lisa-kaltenegger-802/.

25. "Five Snapshots of How the Earth Looked at Key Points in Its History Could Help Us Find Habitable Exoplanets" by M. Williams, 2020, *Universe Today*, https:// www.universetoday.com/145520/five-snapshots-of-how-the-earth-looked-at-key -points-in-its-history-could-help-us-find-habitable-exoplanets/.

26. "Ancient Earth Was a Water World" by P. Voosen, 2021, *Science*, volume 371, 1088–1089.

27. "Five Snapshots of How the Earth Looked" by M. Williams, 2020, *Universe Today*, https://www.universetoday.com/145520/five-snapshots-of-how-the-earth-looked-at -key-points-in-its-history-could-help-us-find-habitable-exoplanets/.

28. "Chance Played a Role in Determining Whether Earth Stayed Habitable" by T. Tyrrell, 2020, *Communications Earth and Environment*, http://dx.doi.org/10.1038 /s43247-020-00057-8.

Chapter 12

1. *Othello* by W. Shakespeare, 1602, act 5, scene 2.

2. "Moons in Depth," NASA Solar System Exploration, https://solarsystem.nasa.gov /moons/in-depth/.

3. Intriguingly, the Moon may have had a brief window of habitability early in its history. About 3.5 billion years ago, volcanic activity and outgassing from the mantle probably led to an atmosphere thicker than that of Mars today, and enough water to form a global layer three millimeters deep. It's unclear if that condition persisted long enough for microbial life to evolve. See "Was There an Early Habit-ability Window for Earth's Moon?" by D. Shultze-Makuch and I. A. Crawford, 2018, *Astrobiology*, volume 18, 1–4.

4. *Satellites of the Outer Planets: Worlds in Their Own Right* by D. A. Rothery, 1999, Oxford: Oxford University Press.

5. "10 Things about Our Solar System's Most Marvelous Moons" by B. Dunford, 2017, NASA Solar System Exploration, https://solarsystem.nasa.gov/news/301/10 -things-about-our-solar-systems-most-marvelous-moons/.

6. "Making a Splash on Mars," 2000, NASA Science, https://science.nasa.gov /science-news/science-at-nasa/2000/ast29jun_1m.

7. "Ocean Worlds in the Outer Solar System" by F. Nimmo and R. T. Pappalardo, 2016, *Journal of Geophysical Research*, volume 121, 1378–1399.

8. "A Common Mass Scaling for Satellite Systems of Gaseous Planets" by R. M. Canup and W. R. Ward, 2006, *Nature*, volume 441, 834–839.

9. "The Habitable Edge of Exomoons" by A. Hadhazy, 2013, *Astrobiology Magazine*, https://www.astrobio.net/meteoritescomets-and-asteroids/the-habitable-edge-of-exomoons/.

10. "Kepler Data Reveals 121 Gas Giants That Could Harbor Habitable Moons" by A. Jorgenson, 2018, *Astronomy Magazine*, https://astronomy.com/news/2018/06/kepler-data-reveals-gas-giants-that-could-harbor-habitable-moons.

11. "Sorry, Star Wars Fans. The Ewoks Are Probably All Still Dead from the Death Star Fallout" by N. T. Redd, 2019, Space.com, https://www.space.com/star-wars-rise-of-skywalker-ewoks-death-star.html.

12. "Can Moons Have Moons?" by J. A. Kollmeier and S. N. Raymond, 2019, *Monthly Notices of the Royal Astronomical Society*, volume 483, L80–L84.

13. "Hunting for Mini-Moons: Exomoons Could Have Satellites of Their Own" by C. Q. Choi, 2018, Space.com, https://www.space.com/42145-exomoons-could-have-own-moons.html.

14. "Loose Ends for the Exomoon Candidate Host Kepler-1625b" by A. Teachey, D. Kipping, C. J. Burke, R. Angus, and A. W. Howard, 2020, *Astronomical Journal*, volume 159, 142–157.

15. "The Habitable Zone for Earth-Like Exomoons Orbiting Kepler-1625b" by D. H. Forgan, 2019, *International Journal of Astrobiology*, volume 18, 510–517.

16. "Europa In Depth," *NASA Solar System Exploration*, https://solarsystem.nasa.gov/moons/jupiter-moons/europa/in-depth/.

17. "Alternative Energy Sources Could Support Life on Europa" by D. Schulze-Makuch and L. N. Irwin, 2001, *Transactions of the American Geophysical Union*, volume 82, 150.

18. "Overlooked Ocean Worlds Fill the Outer Solar System" by J. Wenz, 2017, *Scientific American*, https://www.scientificamerican.com/article/overlooked-ocean-worlds-fill-the-outer-solar-system/.

19. "Ganymede May Harbor Club Sandwich of Oceans and Ice" by W. Calvin, 2014, NASA Jet Propulsion Laboratory, https://www.jpl.nasa.gov/news/ganymede-may-harbor-club-sandwich-of-oceans-and-ice.

20. "Triton's Geyser-Like Plumes: Discovery and Basic Characterization" by L. A. Soderblom et al., 1990, *Science*, volume 250, 410–415.

21. "Welcome to Triton, Neptune's Bizarre Wet Moon That Could Change Where We Look for Alien Life" by J. Carter, 2019, *Forbes*, https://www.forbes.com/sites/jamiecartereurope/2019/09/10/welcome-to-triton-neptunes-bizarre-wet-moon-that-could-change-where-we-look-for-alien-life/.

22. *Chasing New Horizons: Inside the Epic First Mission to Pluto* by A. Stern and D. Grinspoon, 2018, New York: Picador.

23. "Saturn's Titan: A Strict Test for Life's Cosmic Ubiquity" by J. I. Lunine, 2009, *Proceedings of the American Philosophical Society*, volume 153, 403–418.

Chapter 13

1. *Hamlet* by W. Shakespeare, 1603, act 2, scene 2.

2. *Ghost Ship: The Mysterious True Story of the Marie Celeste and Her Missing Crew* by B. Hicks, 2004, New York: Random House Digital.

3. *Troilus and Cressida* by W. Shakespeare, 1609, act 1, scene 3.

4. "Is the Distance from the Earth to the Sun Changing?" by C. Springob, 2015, *Ask an Astronomer*, http://curious.astro.cornell.edu/about-us/41-our-solar-system/the-earth /orbit/83-is-the-distance-from-the-earth-to-the-sun-changing-advanced.

5. "Large-Scale Chaos in the Solar System" by J. Laskar, 1994, *Astronomy and Astrophysics*, volume 287, L9–L12.

6. "Numerical Evidence That the Orbit of Pluto Is Chaotic" by G. J. Sussman and J. Wisdom, 1988, *Science*, volume 241, 433–437.

7. "Existence of Collisional Trajectories of Mercury, Mars, and Venus with the Earth" by J. Laskar and M. Gastineau, 2009, *Nature*, volume 459, 817–819.

8. The idea of mathematical chaos is often misunderstood. It doesn't mean that events are unpredictable from moment to moment, or that nothing is long-term predictable. It just means that complex systems behave in a way that's sensitive to the initial conditions, and a nearly deterministic situation can become chaotic and so hard to project over the long haul. Chaos theory originates with mathematician Henri Poincaré and his study of the three-body problem in 1890. Lorenz applied the idea to weather, and his butterfly example was famously used by Ray Bradbury in his science fiction short story "A Sound of Thunder." For Lorenz's influential paper, see "Deterministic Nonperiodic Flow" by E. N. Lorenz, 1963, *Journal of the Atmospheric Sciences*, volume 20, 130–141.

9. "Chaotic Rhythms of a Dripping Faucet" by R. F. Cahalan, H. Leidecker, and G. D. Cahalan, 1990, *Computers in Physics*, volume 4, 368–382.

10. "Sculpting Our Planetary System" by S. Raymond, 2018, *American Scientist*, https://www.americanscientist.org/article/sculpting-our-planetary-system.

11. "Young Solar System's Fifth Giant Planet?" by D. Nesvorný, 2011, *Astrophysical Journal Letters*, volume 742, 22–28.

12. "How Our Galaxy Will Kill Our Solar System in a Trillion Years, Planet by Planet" by P. Plait, 2020, *Syfy Wire*, https://www.syfy.com/syfywire/how-our-galaxy -will-kill-our-solar-system-in-a-trillion-years-planet-by-planet.

13. "Lens-Like Action of a Star by the Deviation of Light in the Gravitational Field" by A. Einstein, 1936, *Science*, volume 84, 506–507.

14. "Microlensing Searches for Planets" by Y. Tsapras, 2018, *Geosciences*, volume 8, 365–401.

15. "A Terrestrial-Mass Rogue Planet Candidate Detected in the Shortest-Timescale Microlensing Event" by P. Mroz et al., 2020, *Astrophysical Journal*, volume 903, L11–L17.

16. "Unveiling Rogue Planets with NASA Roman Space Telescope" by A. Balzer, 2020, NASA, https://www.nasa.gov/feature/goddard/2020/unveiling-rogue-planets -with-nasas-roman-space-telescope.

17. "ESO Telescopes Help Uncover Large Group of Rogue Planets Yet" by N. Miret-Roig, 2021, ESO, https://www.eso.org/public/news/eso2120/.

18. "Astronomers Discover 1st Possible Exoplanets in Another Galaxy" by P. S. Anderson, 2018, *Earth Sky*, https://earthsky.org/space/exoplanets-distant-galaxy -microlensing-xinyu-dai.

19. "Researchers Say Galaxy May Swarm with Nomad Planets" by A. Freeberg, 2012, Phys.org, https://phys.org/news/2012-02-galaxy-swarm-nomad-planets.html.

20. "Life-Sustaining Planets in Interstellar Space?" by D. J. Stevenson, 1999, *Nature*, volume 400, 32.

21. "Hyper-Velocity and Tidal Stars from Binaries Disrupted by a Massive Black Hole" by J. G. Hills, 1988, *Nature*, volume 331, 687–689.

22. "Hyper-Velocity Stars. III. The Space Density and Ejection History of Main-Sequence Stars from the Galactic Center" by W. R. Brown, M. J. Geller, S. J. Kenyon, M. J. Kurtz, and B. C. Bromley, 2007, *Astrophysical Journal*, volume 671, 1708–1716.

23. "Hypervelocity Planets and Transits around Hypervelocity Stars" by I. Ginsburg, A. Loeb, and G. A. Wegner, 2012, *Monthly Notices of the Royal Astronomical Society*, volume 423, 948–954.

24. "On the Origin of the Near-Infrared Extragalactic Background Light Anisotropy" by M. Zemcov et al., 2014, *Science*, volume 346, 732–735.

25. "Planets Could Travel along with Rogue Hypervelocity Stars, Spreading Life throughout the Universe" by M. Williams, 2014, *Universe Today*, https://www .universetoday.com/116872/planets-could-travel-along-with-rogue-hypervelocity -stars-spreading-life-throughout-the-universe/.

26. *Moby-Dick: The Whale* by H. Melville, 1851, New York: Harper and Brothers, 16.

27. This is a scenario based on the best information we have about how life on Earth began. No star is needed to kick-start life as active geology provides abundant

heat and chemical energy. In some ways, a super-Earth is a more hospitable environment for starting biology than Earth because the larger mass favors tectonic activity and the atmosphere will retain more hydrogen, which is useful as a reducing agent in chemical reactions. Evolution from a simple cell with no nucleus to multicellular organisms and higher-order creatures is extremely uncertain, and may not be inevitable. There are reasonable arguments that the basic mechanisms of biochemistry are universal, but life in an alien physical and chemical environment is unlikely to resemble life on Earth.

28. *The Nature of the Physical World* by A. Eddington, 1928, London: Macmillan. Eddington was also supremely confident in his own abilities. He championed the general theory of relativity soon after it was published by Einstein and was interviewed by a reporter about the theory. The reporter said he had heard that only three people fully understood the theory. Eddington was silent for a while. When the reporter prompted him, he said, "I'm trying to think who the third person is."

Chapter 14

1. "In the Great Silence There Is Great Hope" by N. Bostrom, 2007, commissioned for BBC Radio 3, "The Essay," https://www.nickbostrom.com/papers/fermi.pdf.

2. The exact quote is uncertain. For a discussion, see "'Where Is Everybody?': An Account of Fermi's Question" by E. M. Jones, 1985, Los Alamos Technical Report, archived at the Wayback Machine, http://www.fas.org/sgp/othergov/doe/lanl/la-10311 -ms.pdf.

3. "The Great Filter—Are We Almost Past It?" by R. Hanson, 1998, http://hanson .gmu.edu/greatfilter.html.

4. "In the Great Silence There Is Great Hope" by N. Bostrom, 2007, commissioned for BBC Radio 3, "The Essay," https://www.nickbostrom.com/papers/fermi.pdf.

5. "The Universal Nature of Biochemistry" by N. R. Pace, 2001, *Proceedings of the National Academy of Science*, volume 98, 805–808.

6. *Assembling Life: How Can Life Begin on Earth and Other Habitable Planets?* by D. W. Deamer, 2019, Oxford: Oxford University Press.

7. "Life Is a Highway (of Flying Space Rocks)" by C. S. Powell, 2019, *Discover Magazine*, https://www.discovermagazine.com/the-sciences/life-is-a-highway-of-flying-space-rocks.

8. "Mars Meteorites," NASA Jet Propulsion Laboratory, https://www2.jpl.nasa.gov/snc/.

9. "DNA Damage and Survival Time Course of Deinococcal Cell Pellets during 3 Years of Exposure to Outer Space" by Y. Kawaguchi et al., 2020, *Frontiers of Microbiology*, https://doi.org/10.3389/fmicb.2020.02050.

10. "Could We Really All Be Martians?" by D. Schulze-Makuch, 2021, *Air and Space Magazine*, https://www.airspacemag.com/daily-planet/could-we-really-all-be-martians -180977633/.

11. "Synthetic Biology" by S. A. Benner and A. M. Sismour, 2005, *Nature Reviews Genetics*, volume 6, 533–543.

12. "Decline and Fall of the Martian Empire" by K. Zahnle, 2001, *Nature*, volume 412, 209–213.

13. Lowell was influential in more positive ways. He was the first to locate a telescope at a location chosen to have the sharpest images—a principal now followed for all optical telescopes. He embarked on a search for Planet X, and for the work, hired Clyde Tombaugh, who later discovered Pluto. And he fired his first telescope assistant, Andrew Douglass, for failing to believe his story about canals on Mars. Douglass came to southern Arizona and founded the Steward Observatory, where mirrors for many of the world's largest telescopes have been made and where I work.

14. *The War of the Worlds* by H. G. Wells, 1898, New York: Harper and Brothers, book 1, 12.

15. *Waging War of the Worlds: A History of the 1938 Radio Broadcast and the Resulting Panic* by J. Gosling, 2009, Jefferson, NC: McFarland.

16. *Life on Mars: The Complete Story* by P. Chambers, 1999, London: Blandford.

17. "Every Mission to Mars, Ever" 2020, Planetary Society, https://www.planetary .org/space-missions/every-mars-mission.

18. "Long-Term Drying of Mars by Sequestration of Ocean-Scale Volumes of Water in the Crust" by E. L. Scheller, B. L. Ehlmann, R. Hu, D. J. Adams, and Y. L. Yung, 2021, *Science*, volume 372, 56–62.

19. "Multiple Subglacial Water Bodies below the South Pole of Mars Unveiled by New MARSIS Data" by S. E. Lauro et al., 2020, *Nature Astronomy*, volume 5, 63–70.

20. "Evidence of Life on Mars?" by R. G. Joseph et al., 2019, *Journal of Astrobiology and Space Science Reviews*, volume 1, 40–81.

21. "I'm Convinced We Found Evidence of Life on Mars in the 1970s" by G. V. Levin, 2019, *Scientific American*, https://blogs.scientificamerican.com/observations /im-convinced-we-found-evidence-of-life-on-mars-in-the-1970s/.

22. "NASA Perseverance Rover Finally Scooped Up a Piece of Mars" by N. V. Patel, 2021, *MIT Technology Review*, https://www.technologyreview.com/2021/09/02/1034285/nasa -perseverance-rover-first-mars-sample/.

23. "NASA's Perseverance Rover Finds Organic Chemicals on Mars" by M. Wall, 2021, Space.com, https://www.space.com/nasa-perseverance-rover-organics-mars.

24. "NASA Narrows Design for Rocket to Launch Samples off of Mars" by S. Clark, 2020, *Spaceflight Now*, https://spaceflightnow.com/2020/04/20/nasa-narrows-design -for-rocket-to-launch-samples-off-of-mars/.

25. "An Interview with Dr. Chris McKay" by D. V. Black, 2020, *Spaced-Out Classroom* (blog), https://spacedoutclassroom.com/2020/08/18/an-interview-with-dr-chris-mckay/.

26. "Interview with Chris McKay," in *Talking about Life: Conversations on Astrobiology*, edited by C. D. Impey, 2010, Cambridge: Cambridge University Press, 172.

27. "Interview with Chris McKay, Planetary Scientist of Planetary Systems Branch" by S. Rojo, 2018, NASA, https://www.nasa.gov/content/interview-with-chris-mckay -planetary-scientist-of-planetary-systems-branch.

28. "Possible Life Signs in the Clouds of Venus" by P. S. Anderson, 2021, *Earth Sky*, https://earthsky.org/space/phosphine-disequilibrium-venus-atmosphere-pioneer -venus-1978.

29. "Interview with David Grinspoon," in *Talking about Life: Conversations on Astrobiology*, edited by C. D. Impey, 2010, Cambridge: Cambridge University Press, 183.

30. "NASA Selects 2 Missions to Study 'Lost Habitable' World of Venus" by A. Johnson and K. Fox, 2021, NASA, https://www.nasa.gov/press-release/nasa-selects-2 -missions-to-study-lost-habitable-world-of-venus.

31. "What the Future of Venus Exploration Could Look Like Following Major Discovery" by L. Grush, 2020, *Verge*, https://www.theverge.com/21438514/venus -future-exploration-spacecraft-flagship-missions-nasa-phosphine-detection.

32. "Interview with Carolyn Porco," in *Talking about Life: Conversations on Astrobiology*, edited by C. D. Impey, 2010, Cambridge: Cambridge University Press, 202.

33. "JUICE: Jupiter Icy Moons Explorer," 2021, European Space Agency, https://sci .esa.int/web/juice.

34. "New NASA Mission Will Fly Titan's Frigid Skies to Search for Life's Beginnings" by S. Stirone, 2019, *Scientific American*, https://www.scientificamerican.com/article /new-nasa-mission-will-fly-titans-frigid-skies-to-search-for-lifes-beginnings/.

35. "NASA Announces New Dragonfly Drone Mission to Explore Titan" by D. W. Brown, 2019, *New York Times*, https://www.nytimes.com/2019/06/27/science/nasa -titan-dragonfly-caesar.html.

36. "Returning Samples from Enceladus for Life Detection" by M. Neveu et al., 2020, *Frontiers in Astronomy and Space Sciences*, volume 7, https://doi.org/10.3389/fspas.2020 .00026.

37. "Billionaire Yuri Milner's Breakthrough Initiatives Eyes Private Mission to Seek Alien Life" by M. Wall, 2018, Space.com, https://www.space.com/42384-breakthrough -initiatives-alien-life-search-mission.html.

Chapter 15

1. "Is It Possible to Make a Map of an Exoplanet?" by L. Dartnell, 2019, *Sky at Night Magazine*, https://www.skyatnightmagazine.com/space-science/possible-make-map-of-exoplanet/.

2. "Gaia Warning" by P. Moore, 2005, *High Profiles*, https://highprofiles.info/interview/james-lovelock/.

3. "The Origin of Oxygen in Earth's Atmosphere" by D. Bello, 2009, *Scientific American*, https://www.scientificamerican.com/article/origin-of-oxygen-in-atmosphere/.

4. "Gaia Warning" by P. Moore, 2005, *High Profiles*, https://highprofiles.info/interview/james-lovelock/.

5. *What Is Life* by L. Margulis and D. Sagan, 1999, New York: Simon and Schuster, 28.

6. "Gaia Theory in a Nutshell," Cosmopolis Project, http://www.cosmopolisproject.org/gaia-theory-in-a-nutshell/.

7. Lovelock created a simple computer model called Daisyworld. Daisyworld contains two types of daisies, white and black, that naturally live in a certain temperature range and absorb different levels of heat. If the temperature is low on Daisyworld, the black daisies flourish because they absorb more heat. This causes the planet to warm up. If the temperature is high on Daisyworld, the white daisies flourish and reflect heat back off into space. Even if the luminosity of Daisyworld's sun increases substantially, Daisyworld itself maintains a constant temperature. But the self-regulating aspects do have limits; some changes are too extreme for life to accommodate, with catastrophic consequences. Lovelock saw this as a warning to humans as they tamper with the global climate.

8. "The Oxygenation of the Atmosphere and Oceans" by H. D. Holland, 2006, *Philosophical Transactions of the Royal Society B*, volume 361, 903–915.

9. "Study Warns of False Oxygen Positives in Search for Life on Other Planets," 2021, *Science Daily*, www.sciencedaily.com/releases/2021/04/210413124352.htm.

10. "Disequilibrium Biosignatures over Earth History and Implications for Detecting Exoplanet Life" by J. Krissansen-Totten, S. Olsen, and D. C. Catling, 2018, *Science Advances*, volume 4, 1–13.

11. "Interview with Vikki Meadows," in *Talking about Life: Conversations on Astrobiology*, edited by C. D. Impey, 2010, Cambridge: Cambridge University Press, 295.

12. "Toward a List of Molecules as Potential Biosignature Gases for the Search for Life on Exoplanets and Applications to Terrestrial Biochemistry" by S. Seager, W. Bains, and J. J. Petkowski, 2016, *Astrobiology*, volume 16, 465–485.

13. "The Future of Spectroscopic Life Detection on Exoplanets" by S. Seager, 2014, *Proceedings of the National Academy of Science*, volume 111, 12634–12640.

14. "The Woman Who Might Find Us Another Earth" by C. Jones, 2016, *New York Times Magazine*, https://www.nytimes.com/2016/12/07/magazine/the-world-sees-me-as-the-one-who-will-find-another-earth.html.

15. "The Telescope That Ate Astronomy" by L. Billings, 2010, *Nature*, volume 467, 1028–1030.

16. "Scientific Discovery with the James Webb Space Telescope" by J. Kalirai, 2018, *Contemporary Physics*, volume 59, 251–290.

17. "NASA's Webb Will Seek Atmospheres around Potentially Habitable Exoplanets" by C. Pulliam, 2020, NASA, https://www.nasa.gov/feature/goddard/2020/nasa-s-webb-will-seek-atmospheres-around-potentially-habitable-exoplanets.

18. Chlorophyll uses light in the visible range from 450 nanometers (blue) to 650 nanometers (red), but it is less efficient in using green light around 550 nanometers so that light is reflected and we see plants as green. But a more dramatic change in plant reflectivity occurs in the far red at 700 nanometers, rising from 10 to 80 percent. Plants reflect this infrared energy to avoid overheating during photosynthesis. The effect is a red edge that is a clear biosignature of plants. For more detail, see "Extending the Astrobiological Red Edge" by P. Gilster, 2019, *Centauri Dreams* (blog), https://www.centauri-dreams.org/2019/07/12/extending-the-astrobiological-red-edge/.

19. "Exoplanet Biosignatures: Observational Prospects" by Y. Fujii et al., 2018, *Astrobiology*, volume 18, 739–778.

20. "A Chemical Survey of Exoplanets with ARIEL" by G. Tinetti et al., 2018, *Experimental Astronomy*, volume 46, 135–209.

21. "The Habitable Exoplanet (HabEx) Imaging Mission: Preliminary Science Drivers and Technical Requirements" by B. Mennesson et al., 2016, *Proceedings SPIE 9904, Space Telescopes and Instrumentation*, https://doi: 10.1117/12.2240457.

22. "Pathways to Habitable Worlds," in *Pathways to Discovery in Astronomy and Astrophysics for the 2020s*, by National Academies of Sciences, Engineering, and Medicine, 2021, Washington, DC: National Academies Press.

23. "Direct Multipixel Imaging and Spectroscopy of an Exoplanet with a Solar Gravitational Lens Mission" by L. Hall, 2020, NASA, https://www.nasa.gov/directorates/spacetech/niac/2020_Phase_I_Phase_II/Direct_Multipixel_Imaging_and_Spectroscopy_of_an_Exoplanet/.

24. "Statement of Sara Seager before the House Committee on Science, Space, and Technology," 2013, US House of Representatives, https://docs.house.gov/meetings/SY/SY00/20131204/101546/HHRG-113-SY00-Wstate-SeagerS-20131204.pdf.

Chapter 16

1. Interferometry has been a standard method in radio astronomy for over half a century. The Very Large Array and other interferometers combine the radio waves from many individual telescopes to greatly increase their angular resolution. This works for radio waves because the waves can be combined coherently, without losing crucial phase information, using waveguides from other well-established radio technologies. Optical interferometry works on the same principals, but is much more challenging because optical wavelengths are a million times smaller than radio wavelengths—a few hundred nanometers compared to a few centimeters. Combining the light from different telescopes has only been successfully done over a few tens of meters, such as with the four 8.4-meter telescopes of the Very Large Telescope in Chile.

2. "The Far Future of Exoplanet Direct Characterization" by J. Schneider et al., 2010, *Astrobiology*, volume 10, 121–126.

3. "How Much Did the Apollo Program Cost?," Planetary Society, https://www .planetary.org/space-policy/cost-of-apollo.

4. "Spacecraft Escaping the Solar System" by C. Peat, 2021, Heavens Above, https:// www.heavens-above.com/SolarEscape.aspx.

5. "Exclusive: Humans Placed in Suspended Animation for the First Time" by H. Thomson, 2019, *New Scientist*, https://www.newscientist.com/article/2224004-exclusive -humans-placed-in-suspended-animation-for-the-first-time/.

6. "Starshot," Breakthrough Initiatives, https://breakthroughinitiatives.org/initiative/3.

7. "A Roadmap to Interstellar Flight" by P. Lubin, 2016, *Journal of the British Interplanetary Society*, volume 69, 40–72.

8. "Why Chemical Rockets and Interstellar Travel Don't Mix" by C. A. Scharf, 2017, *Scientific American*, https://blogs.scientificamerican.com/life-unbounded/why -chemical-rockets-and-interstellar-travel-dont-mix/.

9. "Breakthrough Starshot: Mission to Alpha Centauri" by P. Gilster, 2016, *Centauri Dreams* (blog), https://www.centauri-dreams.org/2016/04/12/breakthrough-starshot -mission-to-alpha-centauri/.

10. "Space Lasers and Light Sails: The Tech behind Breakthrough Starshot" by A. Duffy, 2016, *Cosmos Magazine*, https://cosmosmagazine.com/space/space-lasers-and -light-sails-tech-behind-breakthrough-starshot/.

11. "How Will We Receive Signals from Interstellar Probes Like Starshot" by *Universe Today*, 2020, Phys.org, https://phys.org/news/2020-05-interstellar-probes-starshot.html.

12. "Optimized Trajectories to the Nearest Stars Using Lightweight High-Velocity Photon Sails" by R. Heller, M. Hippke, and P. Kervella, 2017, *Astronomical Journal*, volume 154, 115–126.

13. "Challenges," Breakthrough Initiatives, https://breakthroughinitiatives.org/chal lenges/3.

14. "Interstellar Travel—the Wait Calculation and the Incentive Trap of Progress" by A. Kennedy, 2006, *Journal of the British Interplanetary Society*, volume 59, 239–246.

15. "Breakthrough Starshot: Reaching for the Stars" by A. Loeb, 2017, *SciTech Europa Quarterly*, https://lweb.cfa.harvard.edu/~loeb/Loeb_Starshot.pdf.

16. "Inside the Breakthrough Starshot Mission to Alpha Centauri" by A. Finkbeiner, 2016, *Scientific American*, https://www.scientificamerican.com/article/inside-the-break through-starshot-mission-to-alpha-centauri/.

Chapter 17

1. *Wonderful Life: The Burgess Shale and the Nature of History* by S. J. Gould, 1999, New York: W. W. Norton.

2. *Contingency and Convergence: Toward a Cosmic Biology of Body and Mind* by R. Powell, 2020, Cambridge, MA: MIT Press. See also "Contingency, Convergence and Hyper-Astronomical Numbers in Biological Evolution" by A. A. Louis, 2016, *Studies in History and Philosophy of Science Part C*, volume 58, 107–116; "Contingency and Determinism in Evolution: Replaying Life's Tape" by Z. D. Blunt, R. E. Lenski, and J. B. Losos, 2018, *Science*, volume 362, 655.

3. "Would Humans Evolve Again If We Rewound Time?" by J. Horton and T. Taylor, 2019, *BBC Future*, https://www.bbc.com/future/article/20190709-would-humans -evolve-again-if-we-rewound-time.

4. "A Collection of Definitions of Intelligence" by S. Legg and M. Hutter, 2007, in *Proceedings of the 2007 Workshop on Advanced in Artificial General Intelligence: Concepts, Architectures, and Algorithms*, Amsterdam: IOS Press, 17–24.

5. *The Timetables of Technology* by B. Bunch and A. Hellemans, 1993, New York: Simon and Schuster.

6. "Searching for Interstellar Communications" by G. Cocconi and P. Morrison, 1959, *Nature*, volume 184, 844–846.

7. "Attempts to Contact Aliens Date Back More Than 150 Years" by M. Schirber, 2009, Space.com, https://www.space.com/6370-attempts-contact-aliens-date-150-years.html.

8. *Life on Other Worlds: The 20th Century Extraterrestrial Life Debate* by S. J. Dick, 2001, Cambridge: Cambridge University Press.

9. "Frank Drake Is Still Searching for E.T." by C. Tu, 2016, *Science Friday*, https://www.sciencefriday.com/articles/frank-drake-is-still-searching-for-e-t/.

10. "What Is the Water-Hole?," 2003, SETI League, http://www.setileague.org/gen eral/waterhol.htm.

11. "Searching for Good Science—the Cancellation of NASA's SETI Program" by S. J. Garber, 1999, *Journal of the British Interplanetary Society*, volume 52, 3–12.

12. "Hunt for Alien Life Gets Cash Bonanza" by Z. Merali, 2015, *Nature*, volume 523, 392–393.

13. "The Breakthrough Listen Search for Intelligent Life: A 3.95–8.00 GHz Search for Radio Technosignatures in the Restricted Earth Transit Zone" by S. Z. Sheikh et al., 2020, *Astronomical Journal*, volume 160, 29–42.

14. "Towards an All-Sky Radio Telescope for SETI" by S. Croft, 2019, *Astronomy and Geophysics*, volume 60, 22–26.

15. "The Square Kilometer Array (SKA) Radio Telescope: Progress and Technical Directions" by P. J. Hall, R. T. Schilizzi, P. E. F. Dewdney, and T. J. W. Lazio, 2008, *Radio Science Bulletin*, no. 326, 4–19.

16. "Smarter SETI Strategy" by N. Cohen and R. Hohlfeld, 2006, *Sky and Telescope*, https://skyandtelescope.org/astronomy-news/smarter-seti-strategy/.

17. "The Father of SETI: Q&A with Astronomer Frank Drake" by L. David, 2015, Space .com, https://www.space.com/28665-seti-astronomer-frank-drake-interview.html.

18. "Project Cyclops: A Design Study of a System for Detecting Extraterrestrial Intelligent Life" by B. M. Oliver and J. Billingham, NASA Technical Report CR-114445, NASA.

19. "Interview with Jill Tarter," in *Talking about Life: Conversations on Astrobiology*, edited by C. D. Impey, 2010, Cambridge: Cambridge University Press, 306.

20. *Contact* by C. Sagan, 1985, New York: Simon and Schuster.

21. "Interview with Jill Tarter," in *Talking about Life: Conversations on Astrobiology*, edited by C. D. Impey, 2010, Cambridge: Cambridge University Press, 314.

22. "Interview with Seth Shostak," in *Talking about Life: Conversations on Astrobiology*, edited by C. D. Impey, 2010, Cambridge: Cambridge University Press, 322.

23. "What Happens If We Find a Signal?" by S. Shostak, 2021, SETI Institute, https://www.seti.org/what-happens-if-we-find-signal.

24. *The UFO Invasion: The Roswell Incident, Alien Abductions, and Government Cover-ups* by K. Frazier, B. Karr, and J. Nickell, 1997, Amherst, NY: Prometheus Books.

25. "I'm an Astronomer and I Think Aliens Might Be Out There—But UFO Sightings Aren't Persuasive" by C. Impey, 2020, *Conversation*, https://theconversation.com

/im-an-astronomer-and-i-think-aliens-may-be-out-there-but-ufo-sightings-arent
-persuasive-150498.

26. "Preliminary Assessment: Unidentified Aerial Phenomena," 2021, Office of
the Director of National Intelligence, https://www.dni.gov/files/ODNI/documents
/assessments/Prelimary-Assessment-UAP-20210625.pdf.

27. "Why the Military Should Work with Scientists to Study the UFO Phenomenon" by C. D. Impey, 2021, *Military Times*, https://www.militarytimes.com/opinion
/commentary/2021/07/15/why-the-military-should-work-with-scientists-to-study
-the-ufo-phenomenon/.

Chapter 18

1. "How Much SETI Has Been Done? Finding Needles in the n-Dimensional Cosmic
Haystack" by J. T. Wright, S. Kanodia, and E. Lubar, 2018, *Astronomical Journal*,
volume 156, 260–272.

2. "The Evolution of Life in the Universe: Are We Alone?" by J. Tarter, 2007, *Proceedings of the International Astronomical Unions, Volume 2, Highlights of Astronomy*,
Cambridge: Cambridge University Press, 14–29.

3. "NASA and the Search for Technosignatures," 2018, report from the NASA Technosignatures Workshop, NASA, https://arxiv.org/ftp/arxiv/papers/1812/1812.08681
.pdf.

4. "Ice Cores and Climate Change," 2014, British Antarctic Survey, https://www.bas
.ac.uk/data/our-data/publication/ice-cores-and-climate-change/.

5. "Defining the Anthropocene" by S. L. Lewis and M. A. Maslin, 2015, *Nature*,
volume 519, 171–180.

6. "Anthropocene Now: Influential Panel Votes to Recognize Earth's New Epoch"
by M. Subramanian, 2019, *Nature*, https://doi.org/10.1038/d41586-019-01641-5.

7. "Great Acceleration," International Geosphere-Biosphere Programme, http://www
.igbp.net/globalchange/greatacceleration.4.1b8ae20512db692f2a680001630.html.

8. R. Krulwich, "Lucy's Laugh Enlivens the Solar System," 2008, *NPR Morning Edition*,
https://www.npr.org/sections/krulwich/2011/08/05/89700174/lucys-laugh-enlivens
-the-solar-system.

9. "The Silurian Hypothesis: Would It Be Possible to Detect an Industrial Civilization in the Geological Record?" by G. A. Schmidt and A. Frank, 2018, *International
Journal of Astrobiology*, volume 18, 142–150.

10. "Nikolai Kardashev" by L. I. Gurvits, Y. Y. Kovalev, and P. G. Edwards, 2019,
Physics Today, https://doi.org/10.1063.PT.6.4o.20191216a.

11. "Transmission of Information by Extraterrestrial Civilization" by N. S. Kardashev, 1964, *Soviet Astronomy*, volume 8, 217–221.

12. "Earth as a Hybrid Planet: The Anthropocene in an Evolutionary Astrobiological Context" by A. Frank, A. Kleidon, and M. Alberti, 2017, *Anthropocene*, volume 19, 13.

13. "Earth as a Hybrid Planet: The Anthropocene in an Evolutionary Astrobiological Context" by A. Frank, A. Kleidon, and M. Alberti, 2017, *Anthropocene*, volume 19, 18.

14. In fact, the idea first appeared in 1937 in the seminal science fiction novel *Star Maker* by Olaf Stapleton and has since appeared in numerous works of science fiction. Dyson's paper was "Search for Artificial Stellar Sources of Infra-Red Radiation" by F. J. Dyson, 1960, *Science*, volume 131, 1667–1668.

15. "Freeman Dyson, Math Genius Turned Technological Visionary, Dies at 96" by G. Johnson, 2020, *New York Times*, https://www.nytimes.com/2020/02/28/science /freeman-dyson-dead.html.

16. "IRAS-Based Whole-Sky Upper Limits on Dyson Spheres" by R. A. Carrigan, 2009, *Astrophysical Journal*, volume 698, 2075–2086.

17. "The G Search for Extraterrestrial Civilizations with Large Energy Supplies. IV. The Signatures and Information Content of Orbiting Megastructures" by J. T. Wright, K. M. S. Cartier, M. Zhao, D. Jontof-Hutter, and E. B. Ford, 2016, *Astrophysical Journal*, volume 816, 17–39.

18. "Limits from CGRO/EGRET Data on the Use of Antimatter as a Power Source by Extraterrestrial Civilizations" by M. J. Harris, 2002, *Journal of the British Interplanetary Society*, volume 55, 383 393.

19. "Our Sky Now and Then: Searches for Lost Stars and Impossible Effects as Probes of Advance Technological Civilizations" by B. Villarroel, I. Imaz, and J. Bergstedt, 2016, *Astronomical Journal*, volume 152, 76–82. Clarke's "third law" was published in "Clarke's Third Law on UFOs" by A. C. Clarke, 1968, *Science*, volume 159, 255.

20. "What Are the Best Ways to Search for Technosignatures?" by A. Tomaswick, 2021, *Universe Today*, https://www.universetoday.com/150815/what-are-the-best-ways -to-search-for-technosignatures/.

21. "Communications from Superior Galactic Communities" by R. N. Bracewell, 1960, *Nature*, volume 186, 670.

22. "Oumuamua as an N2 Ice Fragment of an Exo-Pluto Surface. II: Generation of Ice Fragments and the Origin of Oumuamua" by S. J. Desch and A. P. Jackson, 2021, *Journal of Geological Research Planets*, volume 126, https://doi.org/10.1029/2020JE006807.

23. "Have We Already Been Visited by Aliens?" by E. Kolbert, 2021, *New Yorker*, https:// www.newyorker.com/magazine/2021/01/25/have-we-already-been-visited-by-aliens.

24. "Aliens Wouldn't Need Warp Drives to Take over an Entire Galaxy, Simulation Suggests" by G. Dvorsky, 2021, *Gizmodo*, https://gizmodo.com/aliens-wouldnt-need -warp-drives-to-take-over-an-entire-1847101242.

Chapter 19

1. This quote has variously been attributed to Arthur C. Clarke, Stanley Kubrick, Carl Sagan, Buckminster Fuller, and a character in Walt Kelly's *Pogo* comic strip. For the full, complicated story, see "Sometimes I Think We're Alone, and Sometimes I Think We're Not. In Either Case, the Idea Is Quite Staggering," 2020, Quote Investigator, https://quoteinvestigator.com/2020/10/22/alone/.

2. "On Modified Human Agents: John Lilly and the Paranoid Style in American Neuroscience" by C. Williams, 2019, *History of the Human Sciences*, volume 32, 84–107.

3. "The Drake Equation Revisited: Part 1" by F. Drake, 2003, *Astrobiology Magazine*, http://www.astrobio.net/alien-life/the-drake-equation-revisited-part-i/.

4. "The Origin of the Drake Equation" by F. Drake and D. Sobel, 2010, *Astronomy Beat*, no. 46, 3.

5. "The Origin of the Drake Equation" by F. Drake and D. Sobel, 2010, *Astronomy Beat*, no. 46, 3.

6. Drake still holds to this reduced form of the equation; his California license plate reads "NEQLSL." It is doubtful many people who encounter him on the road know what that abbreviation means, though he should consider a bumper sticker saying, "Honk if you think we're not alone."

7. "The Present-Day Star Formation Rate of the Milky Way Determined from Spitzer-Detected Young Stellar Objects" by T. P. Robitaille and B. A. Whitney, 2010, *Astrophysical Journal Letters*, volume 710, L11–L15.

8. "One or More Bound Planets per Milky Way Star from Microlensing Observations" by A. Cassan et al., 2011, *Nature*, volume 481, 167–169.

9. "The Galactic Habitable Zone and the Age Distribution of Complex Life in the Milky Way" by C. H. Lineweaver, Y. Fenner, and B. K. Gibson, 2004, *Science*, volume 303, 59–62.

10. "Can SETI Succeed? Not Likely" by E. Mayr, 2010, Planetary Society, http://www.planetary.org/explore/topics/search_for_life/seti/mayr.html.

11. *The Evolution of Complexity by Means of Natural Selection* by J. T. Bonner, 1988, Princeton, NJ: Princeton University Press.

12. "Why E.T. Hasn't Called" by M. Shermer, 2002, *Scientific American*, volume 287, 33.

13. *Lonely Planets: The Natural Philosophy of Alien Life* by D. Grinspoon, 2004, New York: Ecco Press.

14. "The Statistical Drake Equation" by C. Maccone, 2010, *Acta Astronautica*, volume 67, 1366–1383.

15. "The Drake Equation Revisited: An Interview with Sara Seager," 2013, *Astrobiology Magazine*, https://www.astrobio.net/alien-life/the-drake-equation-revisited-an-inter view-with-sara-seager/.

16. "Are We Alone in the Universe? Revisiting the Drake Equation" by L. Sierra, 2016, NASA Exoplanet Exploration, https://exoplanets.nasa.gov/news/1350/are-we -alone-in-the-universe-revisiting-the-drake-equation/.

17. "A New Empirical Constraint on the Prevalence of Technological Species in the Universe" by A. Frank and W. T. Sullivan, 2016, *Astrobiology*, volume 16, 359–362.

Chapter 20

1. The colleagues were Emil Konopinski, Herbert York, and Edward Teller, the architect of the hydrogen bomb. The conversation was painstakingly reconstructed in the 1980s by another physicist, Eric Jones of Los Alamos Lab, long after Fermi had died. Jones sent a series of letters to each of the three men until he had assembled all the pieces of the chronology. See "Where Is Everybody? An Account of Fermi's Question" by E. M. Jones, 1985, Technical Report LA-10311-MS, Los Alamos National Laboratory, US Department of Defense.

2. *Enrico Fermi: Physicist* by E. Segre, 1970, Chicago: Chicago University Press.

3. A classic example is "this statement is false," which is both self-referencing and self-contradictory. There is no inherent contradiction in asking where the intelligent aliens are when you expect them to have visited since there are many plausible reasons why they might exist yet have not visited or made themselves apparent in any way. For the formal definition, see "Paradoxes and Contemporary Logic" by A. Cantini, in *Stanford Encyclopedia of Philosophy*, 2012, https://plato .stanford.edu/.

4. "Paleontological Tests: Human-Like Intelligence Is Not a Convergent Feature of Evolution" by C. H. Lineweaver, in *From Fossils to Astrobiology*, edited by J. Seckbach and M. Walsh, 2008, New York: Springer.

5. "The Case for a Gaian Bottleneck: The Biology of Habitability" by A. Chopra and C. H. Lineweaver, 2016, *Astrobiology*, volume 16, 17–22.

6. "Influence of Supernovae, Gamma Ray Bursts, Solar Flares, and Cosmic Rays on the Terrestrial Environment" by A. Dar, in *Global Catastrophic Risks*, edited by N. Bostrom and M. M. Cirkovic, 2008, Oxford: Oxford University Press.

7. Books have been written on the subject, with dozens of explanations for the Fermi paradox or SETI's Great Silence. See *If the Universe Is Teeming with Aliens . . . Where Is Everybody?: Fifty Solutions to the Fermi Paradox and the Problem of Extraterrestrial Life* by S. Webb, 2002, Göttingen: Copernicus Publications. Or more recently, see *Solving Fermi's Paradox* by D. H. Forgan, 2019, Cambridge: Cambridge University Press.

8. *Bulletin of the Atomic Scientists*, https://thebulletin.org/.

9. Doomsday Clock, *Bulletin of the Atomic Scientists*, https://thebulletin.org/doomsday-clock/current-time/.

10. This "space opera" series of short stories and novels describes an advanced species deploying self-replicating machines to destroy all life on planets where rivals may emerge. Berserkers are named after warriors of Norse legends. The first collection in the series is *Berserker* by B. Saberhagen, 1967, New York: Ballantine Books.

11. Papers presented at the international Search for Extraterrestrial Life and Post-Biological Intelligence symposium, https://meetings.seti.org/Search_Extraterrestrial_Life_Post_Biological_Intelligence.html. See also "Post Post-Biological Evolution" by M. M. Cirkovic, 2018, *Futures*, volume 99, 28–35.

12. *Life on the Mississippi* by M. Twain, 1883, Boston: James R. Osgood and Company, 146.

13. "Dissolving the Fermi Paradox" by A. Sandberg, E. Drexler, and T. Ord, 2018, https://arxiv.org/abs/1806.02404.

14. "The Great Silence: The Controversy concerning Extraterrestrial Intelligent Life" by G. D. Brin, 1983, *Quarterly Journal of the Royal Astronomical Society*, volume 24, 307.

15. "A Message from Earth" by C. Sagan, L. S. Sagan, and F. Drake, 1972, *Science*, volume 175, 881–884.

16. "NASA's Fight to Protect Aliens from Naked Ladies" by M. Hay, 2020, Ozy, https://www.ozy.com/true-and-stories/nasas-fight-to-protect-aliens-from-naked-ladies/273932/.

17. *Murmurs of Earth* by C. Sagan, 1978, New York: Random House. A CD-ROM version was released by Warner New Media in 1992. Both are now out of print, but to celebrate the fortieth anniversary of the record in 2018, a successful Kickstarter campaign by Ozma Records led to a vinyl reissue of the record, which won a Grammy. See "Reprint of NASA's Golden Record Takes Home a Grammy" by C. Cofield, 2018, Space.com, https://www.space.com/39549-voyager-golden-record-reprint-wins-grammy.html. In 2015, NASA uploaded the audio portion of the CD to SoundCloud.

18. "NASA Bans Sex from Outer Space" by N. Wade, 1977, *Science*, volume 197, 1163–1164. The Thomas quote is from his collection of essays *Lives of a Cell*.

19. "Voyager: The Space Explorers That Are Still Going Boldly to the Stars" by D. Campbell and C. Riley, 2012, *Guardian*, https://www.theguardian.com/science/2012/oct/21/voyager-mission-leave-solar-system.

20. "Ten Decisions That Could Shake the World" by M. A. G. Michaud, 2003, *Space Policy*, volume 19, 131–136.

21. "Regarding Messaging to Extraterrestrial Intelligence (METI) / Active Searches for Extraterrestrial Intelligence (Active SETI)," signed by twenty-seven astronomers and Elon Musk, SETI at Home, University of California at Berkeley, https://setiathome.berkeley.edu/meti_statement_0.html.

22. *The Dark Forest* by Liu Cixin, 2008, New York: Tor Books. This is the second in the trilogy *Remembrance of Earth's Past* and a sequel to *The Three-Body Problem*, which won a Hugo Award.

23. "Ode to the Alien" by D. Ackerman, 1979, *Michigan Quarterly Review*, https://sites.lsa.umich.edu/mqr/2017/07/from-the-archive-ode-to-the-alien-by-diane-ackerman/.

Chapter 21

1. "Hadean Earth and Primordial Continents: The Cradle of Prebiotic Life" by M. Santosh, T. Arai, and S. Maruyama, 2017, *Geoscience Frontiers*, volume 8, 309–327.

2. "Snowball Earth" by P. F. Hoffman and D. P. Schrag, 2000, *Scientific American*, http://www.sciam.com/2000/0100issue/0100hoffman.html.

3. *Life on a Young Planet: The First Three Billion Years of Evolution on Earth* by A. H. Knoll, 2004, Princeton, NJ: Princeton University Press.

4. "The Day the Dinosaurs Died" by D. Preston, 2019, *New Yorker*, https://www.newyorker.com/magazine/2019/04/08/the-day-the-dinosaurs-died.

5. *Humans at the End of the Ice Age: The Archaeology of the Pleistocene* edited by L. G. Straus, B. V. Eriksen, J. Erlandson, and D. R. Yesner, 1996, London: Springer.

6. "Brain Evolution, the Determinates of Food Choice, and the Omnivore's Dilemma" by G. J. Armelagos, 2014, *Critical Reviews in Food Science and Nutrition*, volume 54, 1330–1341.

7. "'One Small Step for Man' or 'a Man'" by A. Stamm, 2019, Smithsonian Air and Space Museum, https://airandspace.si.edu/stories/editorial/one-small-step-man-or-man#:~:text=The%20case%20also%20features%20Neil,one%20giant%20leap%20for%20mankind.%22.

8. "Human Population Growth and the Demographic Transition" by J. Bongaarts, 2009, *Philosophical Transactions of the Royal Society B*, volume 364, 2985–2990.

9. "World Hunger: Key Facts and Statistics 2021," Action against Hunger, https://www.actionagainsthunger.org/world-hunger-facts-statistics.

10. "Earth as a Hybrid Planet: The Anthropocene in an Evolutionary Astrobiological Context" by A. Frank, A. Kleidon, and M. Alberti, 2017, *Anthropocene*, volume 19, 13–21.

11. "The Great Acceleration" by Future Earth staff member, 2015, Future Earth, https://futureearth.org/2015/01/16/the-great-acceleration/.

12. "Planetary Boundaries: Guiding Human Development on a Changing Planet" by W. Steffen et al., 2015, *Science*, volume 347, 736.

13. "The Trajectory of the Anthropocene: The Great Acceleration" by W. Steffen, W. Broadgate, L. Deutsch, O. Gaffney, and C. Ludwig, 2015, *Anthropocene Review*, volume 2, 81–98.

14. "The Great Acceleration: An Unequal World Shows the Strain" by Climate News Network, 2015, *Independent Australia*, https://independentaustralia.net/environment/environment-display/the-great-acceleration-unequal-world-shows-the-strain,7278.

15. "Global Wealth Inequality" by G. Zucman, 2019, *Annual Reviews of Economics*, volume 11, 109–138.

16. *World Inequality Report 2022* by L. Chancel et al., 2022, World Inequality Lab, https://wir2022.wid.world/.

17. "The Modern World: The Effect of Democracy, Colonialism and War on Economic Growth 1820–2000" by B. Milanovic, 2005, https://dx.doi.org/10.2139/ssrn.812144; "Globalization, Multiculturalism, and Other Fictions: Colonialism for the New Millenium?" by S. B. Banerjee and S. Linstead, 2001, *Organization*, volume 8, 683–722.

18. "You Bastards Sacked Me. When the Climate Skeptics Arrived" by M. Wilkinson, 2020, *Sydney Morning Herald*, https://www.smh.com.au/national/you-bastards-sacked-me-when-the-climate-sceptics-arrived-20200626-p556nn.html.

19. "Synchronized Peak-Rate Years of Global Resources Use" by R. Seppelt, A. M. Manceur, J. Liu, E. P. Fenichel, and S. Koltz, 2014, *Ecology and Society*, volume 19, 50–64.

20. "Reconsidering the Limits to World Population: Meta-Analysis and Meta-Prediction" by J. C. J. M. Van Den Bergh and P. Rietveld, 2004, *BioScience*, volume 54, 195–204.

21. "Global Human-Made Mass Exceeds All Living Biomass" by E. Elhacham, L. Ben-Uri, J. Grozovski, Y. M. Ban-On, and R. Milo, 2020, *Nature*, volume 588, 442–444.

22. "Human Domination of the Biosphere: Rapid Discharge of the Earth-Space Battery Foretells the Future of Humankind" by J. Schramski, D. K. Gattie, and J. H. Brown, 2015, *Proceedings of the National Academy of Sciences*, volume 112, 9511–9517.

23. "Are We Bigger Than the Biosphere? An Examination of Our Human-Dominated Planet" by Y. Malhi, 2014, School of Geography and the Environment, University of Oxford, https://www.geog.ox.ac.uk/events/150212-ymalhi.pdf.

24. "Metabolic Scaling in Complex Living Systems" by D. S. Glazier, 2014, *Systems*, volume 2, 451–540.

25. "The Surprising Thing I Learned Sailing around the World" by E. MacArthur, 2015, TED Talk transcript, https://www.ted.com/talks/dame_ellen_macarthur_the_surprising_thing_i_learned_sailing_solo_around_the_world/transcript?language=en.

26. "The World's Biggest Membrane," in *Lives of a Cell: Notes of a Biology Watcher*, by L. Thomas, 1974, New York: Viking Press, originally published in *New England Journal of Medicine*, 1973, volume 285, 576.

27. "The Overview Effect: Awe and Self-Transcendent Experience in Space Flight" by D. B. Yaden, J. Iwry, K. J. Slack, J. C. Eichstaedt, Y. Zhao, G. E. Vaillant, and A. B. Newburg, 2016, *Psychology of Consciousness: Theory, Research, and Practice*, volume 3, 1–11.

Chapter 22

1. "The Space Race Is Dominated by New Contenders," 2018, *Economist*, https://www.economist.com/graphic-detail/2018/10/18/the-space-race-is-dominated-by-new-contenders.

2. "A Brief History of the FAA," 2017, Federal Aviation Administration, https://www.faa.gov/about/history/brief_history/.

3. "How Aviation Safety Has Improved," Allianz Global Corporate and Specialty, https://www.agcs.allianz.com/news-and-insights/expert-risk-articles/how-aviation-safety-has-improved.html.

4. "Private Space Companies Avoid FAA Oversight Again, with Congress' Blessing" by L. Grush, 2015, *Verge*, https://www.theverge.com/2015/11/16/9744298/private-space-government-regulation-spacex-asteroid-mining.

5. Earth's atmosphere does not abruptly end, but gradually becomes less dense with altitude until it shades into the pure vacuum of space. The traditional definition of the edge of space, accepted by the Federation Aeronautique Internationale, is 100 kilometers (62 miles, 330,000 feet) above mean sea level. It's named the Kármán line, after a Hungarian engineer and physicist who gave the rationale in 1957, as told in his later autobiography. See *The Wind and Beyond* by T. von Kármán and L. Edson, 1967, New York: Little, Brown and Company, 343. An argument has been

made for a lower boundary of 80 kilometers in "The Edge of Space: Revisiting the Karman Line" by J. C. McDowell, 2018, *Acta Astronautica*, volume 151, 668–677.

6. "How Many Astronauts Have Died in Space?" by J. Parks, 2019, *Astronomy Magazine*, https://astronomy.com/news/2019/10/how-many-astronauts-have-died-in-space.

7. "Your Chance of Dying Ranked by Sport and Activity," 2019, Teton Gravity Research, https://www.tetongravity.com/story/adventure/your-chances-of-dying -ranked-by-sport-and-activity.

8. "Has Anyone Ever Died in Space?" by R. Webb, 2018, *New Scientist*, https://www .newscientist.com/question/anyone-ever-died-space/.

9. "The Nedelin Catastrophe, Part 1" by M. Azriel, 2014, *Space Safety Magazine*, http://www.spacesafetymagazine.com/space-disasters/nedelin-catastrophe/historys -launch-padfailures-nedelin-disaster-part-1/; "The Nedelin Catastrophe, Part 2" by M. Azriel, 2014, *Space Safety Magazine*, http://www.spacesafetymagazine.com/space -disasters/nedelin-catastrophe/failure-launch-pad-nedelin-disaster-part-2/.

10. "Virgin Galactic's SpaceShipTwo Crashes in New Setback for Commercial Space-flight" by K. Chang and J. Schwartz, 2014, *New York Times*, https://www.nytimes.com /2014/11/01/science/virgin-galactics-spaceshiptwo-crashes-during-test-flight.html.

11. "SpaceX Starships Keep Exploding, but It's All Part of Elon Musk's Plan" by E. Olsen, 2021, *Popular Science*, https://www.popsci.com/story/technology/spacex -starship-explosions/.

12. "Elon Musk Tweeted That the Egyptian Pyramids Were Built by Aliens. Here Are 39 of the Most Outrageous Things He's Said over the Years" by M. Matousek and A. Hartmans, 2020, *Business Insider*, https://www.businessinsider.com/elon-musk -shocking-quotes-tweets-2018-10.

13. "Tesla Passes 1 Million EV Milestone and Model 3 Become All Time Best Seller" by M. Holland, 2020, *Clean Technica*, https://cleantechnica.com/2020/03/10/tesla -passes-1-million-ev-milestone-and-model-3-becomes-all-time-best-seller/.

14. "Launch List," 2021, SpaceX Info, https://spacex-info.com/launch-list/

15. "Extended Transcript: SpaceX CEO Elon Musk on Putting Boots on the Moon and Mars" by D. Morgan, 2019, CBS News, https://www.cbsnews.com/news/extended -transcript-spacex-ceo-elon-musk-on-putting-boots-on-the-moon-and-mars/.

16. "Extended Transcipt: SpaceX CEO Elon Musk on Putting Boots on the Moon and Mars" by D. Morgan, 2019, CBS News, https://www.cbsnews.com/news/extended -transcript-spacex-ceo-elon-musk-on-putting-boots-on-the-moon-and-mars/.

17. "Amazon CEO Jess Bezos Championed Invention, Failure, and Customer Obses-sion in His 24 Shareholder Letters. Here Are the 25 Best Quotes" by T. Mohamed,

2021, *Business Insider*, https://markets.businessinsider.com/news/stocks/amazon-ceo-jeff-bezos-25-best-quotes-annual-shareholder-letters-2021-4-1030340504.

18. "Inside Amazon: Wrestling Big Ideas in a Bruising Workplace" by J. Kantor and D. Streitfeld, 2015, *New York Times*, https://www.nytimes.com/2015/08/16/technology/inside-amazon-wrestling-big-ideas-in-a-bruising-workplace.html.

19. "Jeff Bezos' Master Plan" by F. Foer, 2019, *Atlantic Magazine*, https://www.theatlantic.com/magazine/archive/2019/11/what-jeff-bezos-wants/598363/.

20. *The High Frontier: Human Colonies in Space* by G. K. O'Neil, 1977, New York: William Morrow and Company.

21. "Interview: Jeff Bezos Lays Out Blue Origin's Space Vision" by A. Boyle, 2016, *Geek Wire*, https://www.geekwire.com/2016/interview-jeff-bezos/.

22. "Extended Transcript: SpaceX CEO Elon Musk on Putting Boots on the Moon and Mars" by D. Morgan, 2019, CBS News, https://www.cbsnews.com/news/extended-transcript-spacex-ceo-elon-musk-on-putting-boots-on-the-moon-and-mars/.

23. *Losing My Virginity: How I've Survived, Had Fun, and Made a Fortune Doing Business My Way* by R. Branson, 1998, London: Virgin Books; *Screw It, Let's Do It* by R. Branson, 2006, London: Virgin Books. Branson has also written a series of jaunty business books with naughty titles, and one book on his aspirations for adventure and space travel, *Reach for the Skies: Ballooning, Birdmen and Blasting into Space* by R. Branson, 2010, London: Virgin Books.

24. Branson's career was almost derailed before it started. In 1971, the high school dropout had one record store on Oxford Street in London, but he was losing money fast. He cut corners by avoiding paying taxes on vinyl records exported to Europe, and only his mother's intervention prevented him from spending a lot more than one night in jail. He tells the story in his book *Losing My Virginity*. Branson's original company, Virgin Records, has a fond place in British cultural history. Its first big hit was Mike Oldfield's *Tubular Bells*, and Virgin signed controversial bands like the Sex Pistols that other companies spurned and avant-garde music with no obvious mass-market appeal. Later the company would go on sign the Rolling Stones, Peter Gabriel, and Culture Club, and grow to become the world's largest independent record label.

25. "A Retrospective of Burt Rutan's High-Performance Art" by the editors, 2012, *Air and Space Magazine*, https://www.airspacemag.com/flight-today/design-by-rutan-133347555/?no-ist.

26. *Safe Is Not an Option: Overcoming the Futile Obsession with Getting Everybody Back Alive That Is Killing Our Expansion into Space* by R. E. Simberg, 2013, Jackson, WY: Interglobal Media.

27. "NASA and the Rise of Commercial Space: A Symposium to Examine the Meanings and Contexts of Commercial Space," 2021, NASA, https://www.nasa.gov/centers/marshall/history/nasa-and-the-rise-of-commercial-space.html.

28. "2021: More Space Launches Than Any Year in History since Sputnik" by C. D. Impey, 2021, *Hill*, https://thehill.com/opinion/technology/587630-2021-more-space-launches-than-any-year-in-history-since-sputnik.

Chapter 23

1. "Konstantin Tsiolkovsky: Russian Father of Rocketry" by N. T. Redd, 2013, Space.com, https://www.space.com/19994-konstantin-tsiolkovsky.html.

2. *Beyond: Our Future in Space* by C. D. Impey, 2016, New York: W. W. Norton.

3. "Budget Documents, Strategic Plans and Performance Reports," 2021, NASA, https://www.nasa.gov/news/budget/index.html.

4. "A Short History of the Internet," 2020, Science and Media Museum, https://www.scienceandmediamuseum.org.uk/objects-and-stories/short-history-internet.

5. "The World's Technological Capacity to Store, Communicate, and Compute Information" by M. Hilbert and P. Lopez, 2011, *Science*, volume 332, 60–65.

6. "Measuring the U.S. Internet Sector: 2019" by C. Hooton, 2019, Internet Association, https://internetassociation.org/publications/measuring-us-internet-sector-2019/.

7. "Bank of America Expects the Space Industry to Triple to a $1.4 Trillion Market within a Decade" by M. Sheetz, 2020, CNBC, https://www.cnbc.com/2020/10/02/why-the-space-industry-may-triple-to-1point4-trillion-by-2030.html.

8. "Exponential Laws of Computing Growth" by P. J. Denning and T. G. Lewis, 2017, *Communications of the Association for Computing Machinery*, volume 60, 54–65.

9. The rocket equation as formulated by Tsiolkovsky is delta $V = X \ln(M_i/M_f)$, where delta V is the change in velocity of the rocket, X is the exhaust velocity, M_i is the initial weight of the rocket, and M_f is the final weight of the rocket. The logarithm is a natural logarithm to the base e (Euler's constant). The fact that the weight ratio is inside a logarithm is the reason that the gain in velocity is so modest as the mass of the rocket (mostly fuel) increases. For a clear and nontechnical description, see "The Tyranny of the Rocket Equation" by D. Pettit, 2012, NASA, https://www.nasa.gov/mission_pages/station/expeditions/expedition30/tryanny.html.

10. *Day Dreams of Heaven and Earth* by K. E. Tsiolkovsky, 1895, Moscow, Russia: USSR Academy of Sciences.

11. *The Fountains of Paradise* by A. C. Clarke, 1979, New York: Harcourt Brace Jovanovich; "The Space Elevator: Thought Experiment, or Key to the Universe" by A. C.

Clarke, 1981, *Advances in Earth Oriented Applied Space Technologies*, volume 1, 39–48, reprinted online in 2003, SpaceRef, http://spaceref.com/space-elevator/the-space -elevator-thought-experiment-or-key-to-the-universe-by-sir-arthur-c-clarke.html. Clarke's book won both a Hugo Award and Nebula Award for best novel.

12. *Ascent to Orbit: A Scientific Autobiography. The Technical Writings of Arthur C. Clarke* by A. C. Clarke, 1984, New York: Wiley, 193.

13. "Audacious and Outrageous: Space Elevators" by S. Price, 2000, NASA Science, https://science.nasa.gov/science-news/science-at-nasa/2000/ast07sep_1.

14. *Road to the Space Elevator Era* edited by P. A. Swan, D. I. Raitt, J. M. Knapman, A. Tsuchida, and M. A. Fitzgerald, 2019, *International Academy of Astronautics*, Heinlein ebook.

15. "This Building Hanging from an Asteroid Is Absurd—But Let's Take It Seriously for a Second" by S. Chodosh, 2017, *Popular Science*, https://www.popsci.com /building-hanging-from-an-asteroid/.

16. "Launch Costs to Low Earth Orbit, 1980–2100," 2018, Future Timeline, https:// www.futuretimeline.net/data-trends/6.htm.

17. "The Orbital Tower: A Spacecraft Launcher Using the Earth's Rotational Energy" by J. Pearson, 1975, *Acta Astronautica*, volume 2, 785–799.

18. "Space Elevators: The Green Road to Space" by P. Swan, C. Swan, P. Phister, D. Dotson, J. Bernard-Cooper, and B. Molloy, 2021, ISEC Position Paper 2021–1, https://www.isec.org/studies/#GreenRoad.

19. *The Space Elevator: A Revolutionary East-to-Space Transportation System* by B. C. Edwards and A. C. Westling, 2003, published by BC Edwards; *Leaving the Planet by Space Elevator* by B. C. Edwards and P. Ragan, 2010, published by lulu.com.

20. The entire arc of this chapter, and this part of the book, tells a story that is propelled by the world's richest and most powerful country. The restless entrepreneurial spirit that motivates the space pioneers is a quintessential US attribute. It's worth stepping back and noting how specifically this history is rooted in the Western scientific tradition. My acquaintance with Eastern cultures leads me to speculate that if the world's dominant culture and religion had been Buddhism, with its emphasis on self-knowledge, compassion, and the alleviation of suffering, we might never have gone into space at all.

Chapter 24

1. "Space Debris and Human Spacecraft" by M. Garcia, 2021, NASA, https://www .nasa.gov/mission_pages/station/news/orbital_debris.html.

2. "The Kessler Syndrome" by D. J. Kessler, 2009, Charter.net, https://web.archive .org/web/20100527195029/http://webpages.charter.net/dkessler/files/KesSym.html.

3. "Space Junk: How Cluttered Is the Final Frontier?" by E. Kwong, 2020, National Public Radio, https://www.npr.org/2020/01/10/795246131/space-junk-how-cluttered -is-the-final-frontier.

4. "Space Junk Is Our New Tragedy of the Commons" by A. Kluth, 2021, *Bloomberg*, https://www.bloomberg.com/opinion/articles/2021-04-17/space-junk-like-overfishing -and-pollution-is-a-global-tragedy-of-the-commons.

5. "Space Ethics according to Space Ethicists" by J. S. J. Schwartz and T. Mulligan, 2021, *Space Review*, https://www.thespacereview.com/article/4117/1.

6. "Why We Need to Stop Talking about Space as a Frontier" by L. Messeri, 2017, *Slate Magazine*, https://slate.com/technology/2017/03/why-we-need-to-stop-talking -about-space-as-a-frontier.html.

7. "Ethical Exploration and the Role of Planetary Protection in Disrupting Colonial Practices" by F. Tavares et al., 2020, submission to the Planetary Science and Astrobiology Decadal Survey 2023–2032, https://arxiv.org/ftp/arxiv/papers/2010/2010.08344 .pdf.

8. "To Boldly Go . . . Responsibly" by J. Sercel and J. Kwast, 2020, *Politico*, https:// www.politico.com/news/2020/11/20/space-ethics-opinion-438526.

9. *The Ethics of Space Exploration* edited by J. S. J. Schwartz and T. Milligan, 2016, New York: Springer.

10. "A Planetary Park System for Mars" by C. Cockell and G. Horneck, 2004, *Space Policy*, volume 20, 291–295.

11. "Is Mars Ours?" by A. Mann, 2021, *New Yorker*, https://www.newyorker.com /science/elements/is-mars-ours.

12. "The Artemis Accords: Principles for Cooperation in the Civil Exploration and Use of the Moon, Mars, Comets, and Asteroids for Peaceful Purposes," NASA, https://www.nasa.gov/specials/artemis-accords/img/Artemis-Accords-signed -13Oct2020.pdf.

13. "Resolution Adopted by the General Assembly, Treaty on Principles Governing the Activities of States in the Exploration and Use of Outer Space, Including the Moon and Other Celestial Bodies," 1965, United Nations Office for Outer Space Affairs, https:// www.unoosa.org/oosa/en/ourwork/spacelaw/treaties/outerspacetreaty.html.

14. "Agreement Governing the Activities of States on the Moon and Other Celestial Bodies," 1979, United Nations Office for Outer Space Affairs, https://www.unoosa .org/oosa/en/ourwork/spacelaw/treaties/intromoon-agreement.html.

15. "Artemis Accords: Why Many Countries Are Refusing to Sign Moon Exploration Agreement" by C. Newman, 2020, *Conversation*, https://theconversation.com

/artemis-accords-why-many-countries-are-refusing-to-sign-moon-exploration
-agreement-148134.

16. "Russia, Once a Space Superpower, Turns to China for Missions" by A. E. Kramer and S. L. Myers, 2021, *New York Times*, https://www.nytimes.com/2021/06 /15/world/asia/china-russia-space.html.

17. *Mission to Mars: My Vision for Space Exploration* by B. Aldrin and L. David, 2013, Washington, DC: National Geographic Books.

18. "Elon Musk Reminds Us All That a Bunch of People Will Probably Die Going to Mars" by C. Gohd, 2021, Space.com, https://www.space.com/elon-musk-mars -spacex-risks-astronauts-die.

19. "Elon Musk Is Going to Do This Mars Thing" by M. Koren, 2021, *Atlantic*, https://www.theatlantic.com/science/archive/2021/05/elon-musk-spacex-starship -launch/618781/.

20. *Starship User's Guide*, 2020, Space Exploration Technologies Corporation, https:// www.spacex.com/media/starship_users_guide_v1.pdf.

21. "The SpaceX Starship Is a Very Big Deal" by C. Handmer, 2019, *Casey Handmer's Blog*, https://caseyhandmer.wordpress.com/2019/10/29/the-spacex-starship-is-a-very -big-deal/.

22. "Moon Hole Might Be Suitable for Colony," 2010, CNN, http://www.cnn.com /2010/TECH/space/01/01/moon.lava.hole/.

23. "3D Printing Our Way to the Moon," 2019, European Space Agency, https://www .esa.int/Enabling_Support/Preparing_for_the_Future/Discovery_and_Preparation /3D_printing_our_way_to_the_Moon.

24. "NASA Inches Closer to Printing Artificial Organs in Space" by T. Woodall, 2021, *MIT Technology Review*, https://www.technologyreview.com/2021/06/18/1026556/nasa -bioprinting-artificial-organs-space/.

25. "How to Feed a Mars Colony of 1 Million People" by C. Q. Choi, 2019, Space .com, https://www.space.com/how-feed-one-million-mars-colonists.html.

Chapter 25

1. "Build the Economy Here on Earth by Exploring Space: Tyson" by K. Kramer, 2017, CNBC, https://www.cnbc.com/2015/05/01/build-the-economy-here-on-earth -by-exploring-space-tyson.html.

2. "We're Never Going to Mine the Asteroid Belt" by D. Fickling, 2020, *Bloomberg*, https://www.bloomberg.com/opinion/articles/2020-12-21/space-mining-on-asteroids -is-never-going-to-happen.

3. "Earth Overshoot Days Falls on July 29," 2021, Earth Overshoot Day, https://www.overshootday.org/.

4. *The Race for What's Left* by M. T. Klare, 2012, New York: Metropolitan Books.

5. "The End of the Oil Age," 2003, *Economist*, https://www.economist.com/node/2155717/print.

6. "Mineral Resources: Reserves, Peak Production and the Future" by L. D. Meinart, G. R. Robinson Jr., and N. T. Nassar, 2016, *Resources*, volume 6, 14–28.

7. *Mining the Sky: Untold Riches from the Asteroids, Comets, and Planets* by J. S. Lewis, 1998, New York: Basic Books.

8. "Don't Panic about Rare Earth Elements" by J. Hsu, 2019, *Scientific American*, https://www.scientificamerican.com/article/dont-panic-about-rare-earth-elements/.

9. "Tons of Water in Asteroids Could Fuel Satellites, Space Exploration" by M. T. Redd, 2019, Space.com, https://www.space.com/water-rich-asteroids-space-exploration-fuel.html.

10. "How Many Ore-Bearing Asteroids?" by M. Elvis, 2013, *Planetary and Space Science*, volume 91, 20–26.

11. "Asteroid Database and Mining Rankings" by I. Webster, 2021, Asterank, http://www.asterank.com/.

12. "NASA Finds Metal Asteroid Worth More Than the Global Economy" by C. Jamasmie, 2020, Mining.com, https://www.mining.com/nasa-finds-rare-metal-asteroid-worth-more-than-global-economy/.

13. "Space Law Treaties and Principles," 2021, United Nations Office for Outer Space Affairs, https://www.unoosa.org/oosa/en/ourwork/spacelaw/treaties.html.

14. Another important treaty, the Partial Test Ban Treaty, was signed in 1963. It was spurred by widespread concern over the testing of nuclear weapons in the atmosphere and the worry that those weapons could be deployed in space. The treaty prohibits the testing of all nuclear weapons except underground. It was signed by 123 nations, including the world's major nuclear powers. The miniaturization of nuclear weapons and the secretive nature of the Chinese space program has led to concerns that the treaty may be difficult to enforce.

15. "Agreement Governing the Activities of States on the Moon and Other Celestial Bodies," 2021, United Nations Office for Outer Space Affairs, https://www.unoosa.org/oosa/en/ourwork/spacelaw/treaties/moon-agreement.html.

16. "Asteroid Mining Made Legal after Passing of Historic Space Bill by US" by M. Molloy, 2015, *Telegraph*, https://www.telegraph.co.uk/news/worldnews/northameri

ca/usa/12019740/Who-owns-space-Asteroid-mining-made-legal-in-US-after-passing
-of-2015-space-bill.html.

17. "Space Law Is Inadequate for the Boom in Human Activity There," 2019, *Economist*, https://www.economist.com/international/2019/07/18/space-law-is-inadequate
-for-the-boom-in-human-activity-there.

18. "In Space, No One Can Hear You Mine: NASA's Blueprint for Space Mining" by
M. Hall, 2021, Mining Technology, https://www.mining-technology.com/features
/in-space-no-one-can-hear-you-mine-nasas-blueprint-for-space-mining/.

19. "One Giant Leap for Capitalistkind: Private Enterprise in Outer Space" by
V. L. Shammas and T. B. Holen, 2019, *Humanities and Social Sciences Communications*,
volume 5, 10–19.

20. "What Is NASA's Asteroid Redirect Mission?" by J. Wilson, 2018, NASA, https://
www.nasa.gov/content/what-is-nasa-s-asteroid-redirect-mission.

21. "New Class of Easily Retrievable Asteroids That Could Be Captured with Rocket
Technology Found" by K. Mohan, 2013, *International Business Times*, https://www
.ibtimes.com/new-class-easily-retrievable-asteroids-could-be-captured-rocket
-technology-found-1382529.

22. "Cosmic Collisions and the Longevity of Non-Spacefaring Galactic Civilizations" by S. J. Ostro and C. Sagan, 1998, NASA Jet Propulsion Laboratory, https://trs
.jpl.nasa.gov/bitstream/handle/2014/19498/98-0908.pdf.

23. *Asteroid Retrieval Feasibility Study* by J. Brophy et al., 2012, Keck Institute for
Space Studies, https://kiss.caltech.edu/final_reports/Asteroid_final_report.pdf.

24. "Optical Mining," 2021, TransAstra Corporation, https://www.transastracorp
.com/optical-mining.

25. "Space Station Biomining Experiment Demonstrates Rare Earth Element Extraction in Microgravity and Mars Gravity" by C. S. Cockell et al., 2020, *Nature Communications*, volume 11, 5523–5534.

26. "How the Asteroid-Mining Bubble Burst" by A. A. Abrahamian, 2019, *MIT Technology Review*, https://www.technologyreview.com/2019/06/26/134510/asteroid-mining
-bubble-burst-history/.

27. "Asteroid Mining Venture Could Change Supply/Demand Ratio on Earth" by
P. Suciu, 2012, *Red Orbit*, https://www.redorbit.com/news/space/1112523850/asteroid
-mining-venture-could-change-supplydemand-ratio-on-earth/.

Chapter 26

1. The hero's journey in mythology was popularized by writer Joseph Campbell, and permeated the thinking of psychiatrist Carl Jung and philosopher Friedrich Nietzsche. It continues to resonate as it gets used in many books and movies. Variations of the hero's journey can be seen in stories from the Bible as well as books like *Lord of the Rings, Pinocchio, The Little Prince,* and the *Harry Potter* series. In a story told about the Buddha, someone told him that the knowledge and wisdom he gained could also be obtained without ever leaving his home or family. He replied, "Yes, but to understand that I first had to leave my home."

2. *The Time Machine* by H. G. Wells, 1895, London: William Heinemann.

3. *The Dispossessed* by U. K. Le Guin, 1974, New York: Harper and Row.

4. *Parable of the Sower* by O. E. Butler, 1993, New York: Four Walls Eight Windows.

5. "Urban Development Sector," 2021, European Investment Bank, https://www.eib.org/en/projects/sectors/urban-development/index.htm.

6. "Self-Contained Cities: Hyperdense Arcologies of Urban Fantasy and Utopian Fiction" by K. Kohlstedt, 2018, *99 Percent Invisible*, https://99percentinvisible.org/article/self-contained-cities-hyperdense-arcologies-urban-fiction-utopian-fantasy/. For a vision of how cities might transform themselves in a hundred years, see "Ectopia 2121: Super Ecofriendly Cities of the Future," 2021, Ectopia 2121, https://www.ecotopia2121.com/.

7. "What Abu Dhabi's City of the Future Looks Like Now" by A. Flint, 2020, *Bloomberg*, https://www.bloomberg.com/news/articles/2020-02-14/the-reality-of-abu-dhabi-s-unfinished-utopia.

8. "Closed Ecological Systems: Can They Save the Future?" by L. Flynn, 2019, *Money Inc.*, https://moneyinc.com/closed-ecological-systems-can-they-save-the-future/.

9. "Space Settlements: A Design Study" edited by R. D. Johnson and C. Holbrow, 1977, NASA Scientific and Technical Office, SP-413, https://archive.org/details/SpaceSettlementsADesignStudy1977/page/n1/mode/2up.

10. Mars City Design, https://www.marscitydesign.com/.

11. "Urban Planning in Space: 3 Off-World Designs for Future Cities" by M. Colagrossi, 2020, *Big Think*, https://bigthink.com/alchemist-city/urban-planning-in-space/.

12. "Closed Ecological Systems" by C. Tamponnet and C. Savage, 1994, *Journal of Biological Education*, volume 28, 167–174.

13. *2312* by K. S. Robinson, 2012, London: Orbit Books.

14. "Can Science Fiction Wake Us Up to Our Climate Reality?" by J. Rothman, 2022, *New Yorker*, https://www.newyorker.com/magazine/2022/01/31/can-science-fiction-wake-us-up-to-our-climate-reality-kim-stanley-robinson.

15. "Life-Centered Ethics and the Human future in Space" by M. N. Mautner, 2017, *Bioethics*, volume 23, 433.

16. "Gerard K. O'Neil on Space Colonies" by M. Davis, 1974, *Omni Magazine*, https://omnimagazine.com/interview-gerard-k-oneill-space-colonies/.

17. "Gerard K. O'Neil on Space Colonies" by M. Davis, 1974, *Omni Magazine*, https://omnimagazine.com/interview-gerard-k-oneill-space-colonies/.

18. "The Colonization of Space" by G. K. O'Neil, 1974, *Physics Today*, volume 27, 36.

19. "Jeff Bezos Unveils Blue Origin's Vision for Space, and a Moon Lander" by K. Chang, 2019, *New York Times*, https://www.nytimes.com/2019/05/09/science/jeff-bezos-moon.html.

20. "Elon Musk Elaborates on His Proposal to Nuke Mars" by R. Grush, 2015, *Verge*, https://www.theverge.com/2015/10/2/9441029/elon-musk-mars-nuclear-bomb-colbert-interview-explained.

21. "Planting an Ecosystem on Mars" by L. Hall, 2015, NASA Mars Exploration Program, https://mars.nasa.gov/news/1811/planting-an-ecosystem-on-mars/.

22. "Can We Make Mars Earth-Like through Terraforming?" by J. Mehta, 2021, Planetary Society, https://www.planetary.org/articles/can-we-make-mars-earth-like-through-terraforming.

23. *Editing Humanity: The CRISPR Revolution and the New Era of Genome Editing* by K. Davies, 2020, Cambridge, UK: Pegasus Publishers.

24. "My Genome" by Veritas, 2021, https://www.veritasgenetics.com/mygenome/.

25. "Engineering the Perfect Astronaut" by A. Regalado, 2017, *MIT Technology Review*, https://www.technologyreview.com/2017/04/15/152545/engineering-the-perfect-astronaut/.

26. "The Explorer Gene: Is Adventure in Your Blood?," 2017, Bushgear, https://www.bushgear.co.uk/blogs/bush-telegraph-2017/the-explorer-gene-is-adventure-in-your-blood.

27. "How Can Humans Survive Longer in Space? Photosynthetic Skin" by C. Mason, 2021, Science Friday, https://www.sciencefriday.com/articles/chlorohumans-space/, excerpted from *The Next 500 Years: Engineering Life to Reach New Worlds* by C. Mason, 2021, Cambridge, MA: MIT Press.

28. *Traveler's Guide to the Stars* by L. Johnson, 2022, Princeton, NJ: Princeton University Press.

29. *Citadelle* by A. de Saint-Exupéry, 2000, Paris: Editions Gallimard, section 75, 687.

Index

Page numbers in italics refer to figures.